Great Webinars

Interactive learning that is captivating,

informative and fun

Revised and Updated

"Learning world, pay attention. This book is a gift to our field. It is written in a smart, authentic, practical, sassy, and easy to follow way. Great Webinars reminds us to bring our participants' experience front and center if we want them to be enthralled, and the learning environment to be enriched. Cynthia Clay lets her own students tell us what is wrong with the current use of webinars and then proceeds to teach us how to build 'em better. It's like having her sit beside you saying, 'Don't worry... I'll get you there.' And she does. My hat is off."

Beverly Kaye, Founder: Career Systems International
Co-author: Love 'Em or Lose 'Em:
Getting Good People to Stay

"In the Middle Ages, one apprentice worked with one Master for seven years in order to learn his craft. Few of us enjoy the luxury of such leisurely learning. The pace of change in the federal workforce dictates that every agency must find ways to greatly accelerate the pace of learning while preserving the level of interaction, the personal accountability for learning, and the quality of peer-to-peer dialog that we most value in classroom-based programs. Great Webinars *provides a profoundly practical and easy-to-follow template for creating and facilitating stimulating webinars that engage learners as active participants, while creating the kind of energetic 'buzz' that is the hallmark of successful learning experiences. The book is sure to benefit both seasoned instructors and subject matter experts who are new to teaching."*

~ Carol Willett, former Chief Learning Officer,
US Government Accountability Office

"When I finished the virtual facilitation course, I had everything I needed to design and deliver my own successful web workshops, and it is all captured in this practical, detailed book. Thanks to this training and inspiration, I am launching my first online workshop next month. This guide contains everything a training professional needs to create and facilitate great classes online. It's terrific!"

~ Lynn Gaertner-Johnston, founder of Syntax Training

"I have worked with Netspeed Learning Solutions at my last two companies. Cynthia is a pioneer in virtual learning and her workshops are engaging, relevant and productive. In Great Webinars, *she shares tips and techniques that elevate the virtual classroom experience".*

~ Jean Foster
Director of Client Operations and Project Management,
StrutuCore

"The form of virtual learning presented in this book – not just webinars but all the interactive tools such as polling, chat, whiteboards, and collaboration – is how training professionals will be training from here on out. In her course and with this book, Cynthia has set the expectation of what a professional virtual trainer should be able to do. As training professionals, it is imperative that we take this approach seriously, and that we master it with the same level of professionalism as when we mastered classroom facilitation."

~ Jeanette Nyden, founder and president of J. Nyden & Co.

"I've had the pleasure of working with Cynthia and her team at Netspeed Learning Solutions on and off for about 18 years. As a training partner, she realized early and quickly that our biggest challenge, as well as opportunity in the learning space, was to use emerging technology tools to bring us together, rather than to isolate us. She and her team lead in the virtual learning space, and teach those very skills to the rest of us in an engaging and interactive process that allows us to not just 'see' but experience what success looks like. You need this book even if you are experienced with virtual delivery, to refresh and review, and to keep you at the highly connected and winning edge of your virtual sessions."

~ Tiffany Austin,
Organization Effectiveness Consultant,
a leading insurance company

"Cynthia has taken a complex topic and provided a step-by-step plan for successfully repurposing training for the virtual environment. The tools and samples she provides are a testament to her vast experience in the 'repurposing' arena. We can all learn from her successful efforts."

~ Dan Roberts, President of Ouellette & Associates and
Contributing Author, Leading IT Transformation

"Darn it! I can no longer multi-task when trainers have taken Cynthia Clay's strategies, tools, techniques to heart. I am too involved. Cynthia has created a new bible for virtual instruction. Great Webinars *provides the tools, techniques and strategies to transform your professional approach to virtual training forever. It is a must read for any trainer who wants to create outstanding virtual learning experiences."*

~ Mary McGlynn,
President/Partner, PowerSpeaking, Inc.

"Cynthia Clay, Master of Webinars, is at it again and we get to enjoy the results of her new ideas in this update of Great Webinars. Her interactive style is out of this world engaging, effective, and creative, leaving participants with no choice but to learn, apply, and have a ton of fun while doing it!"

~ Kassy LaBorie
Principal Consultant, Kassy LaBorie Consulting, LLC
Author "Interact and Engage! 50+ Activities for Virtual Training, Meetings, and Webinars"

"The wave of the future is how to deliver great content in a fun and engaging way without being in the classroom. Here is the book that will show you how to do just that! Avoid the pitfalls that cause so many online programs and webinars to fail. A must read for those who expect to be a global business!"

~ Anne Warfield,
CSP & Outcome Strategist,
Impression Management Professionals

"After having first participated in the "Virtual Facilitator" program with Cynthia Clay, then reading her book, I completely changed my attitude regarding the validity of a virtual learning classroom. Not only has Cindy put the adult learning principles into practice, she has created an approach to the design, development and delivery of a virtual workshop that is potentially more engaging than being face-to-face. Great Webinars captures all the essentials to creating a great on-line learning experience."

~ Karen Pacent,
Former Director, Learning & Leadership Development
United States Tennis Association

"Cindy and NetSpeed Learning Solutions are the absolute best in getting companies set up for great, interactive webinars. We've been working with Cindy for over 10 years now and she's taught us everything we know about designing and facilitating anything virtual."

~ Fred Allemann
Director of Organization Development
United States Tennis Association

Great Webinars,
Revised and Updated

Interactive learning that is
captivating, informative and fun

Cynthia Clay

Punchy Publishing
Seattle, Washington

International Standard Book Number: 978-0-9764587-6-0

First printing March 2019

Cover Design by Mao Studios

Punchy Publishing
Seattle, WA 98115

http://netspeedlearning.com/

Dedication

For my father, Jack Clay, who has lovingly supported me and my work for many years.

For my husband, Leo Brodie, who continues to inspire, challenge, and support me every day.

And for my daughters, Grayce and Jessica, who have motivated me to explore the neuroscience of learning.

Contents

Foreword

I met Cynthia Clay through the Seattle Chapter of the Women Presidents Organization (WPO), where I had come to know and respect her as a talented business owner. Because her basic approach to leadership matched my own, I enrolled our management team into her NetSpeed Leadership program. Our intent was to bring our leadership team into our corporate headquarters once a month for a full day program with one of the NetSpeed facilitators.

Somewhat serendipitously, it was at about this time that Cynthia invited our team to attend her free marketing webinars, which she had designed to expose her prospects to some of the basic NetSpeed Leadership concepts. It was there that I recognized not only her authority on a broad range of leadership practices, but that she possessed the special expertise to apply adult learning principles in an online environment. Our team quickly scrapped the "all day training session" format, preferring instead her online format.

Meanwhile, in our WPO discussions, Cynthia was sharing her business challenges of getting content out to her customers in an economic climate where it was becoming increasingly more difficult to cost justify having large numbers of people away from their desks for long periods of time (typically including travel time) "for training." While the demand for leadership training was increasing, so was the need to cut costs. The world wanted more training for less money.

My own experience with distributive learning/ communication environments paralleled Cindy's business challenge. My organization, the Pace Staffing Network, had five offices separated by over 100 miles across the Puget Sound region. Since our inception, we had faced the challenge of maintaining good communication with our employees and customers when senior leadership couldn't be physically present. We were, therefore, one of the first to jump into the exploration of online meeting and webinar platforms, because we valued being exposed to ideas from colleagues and partners throughout the world while working at our desks.

When I saw what Cynthia was doing, in her mind from a "marketing" framework, I just put two and two together. Cynthia was a personally talented presenter who not only understood the content of leadership development curricula, but had mastered all the engaging elements of face-to-to face training that drives successful learning. I'm proud to say, as a result of my encouragement, Cynthia took my advice and has established herself as a "guru" in the webinar business, taking the basic concepts to a whole new level of performance.

It's truly been a pleasure watching Cynthia become recognized as an expert in blending virtual delivery with social collaboration and microlearning. I was thrilled to see that she was featured as a thought leader in TD magazine (October 2017). And I was delighted to hear that she has delivered well-attended sessions at the Association for Talent Development (ATD) International Conference for the past 12 years, talking about the very tools that she explores in this book.

Today, she doesn't just deliver webinars, she delivers *extraordinary webinars*. She generously has shared her expertise with companies like the Pace Staffing Network who want to deliver high-quality webinars to our own clients. When Cindy learned that we would be doing webinars for our customers, she offered to coach and train us, encouraged us to purchase the software, and taught us how to use the webinar environment to help solidify our customer relations. She made it really simple for us to look really good in the webinar delivery business.

Now, with her updated and revised *Great Webinars* book and her Virtual Facilitator training courses, she's making her wisdom and experience available to consultants, small businesses, trainers in larger organizations and anyone who wants to deliver engaging, interactive training via web workshops. She has managed to capture the key elements of learning interaction, and translated them to the virtual environment.

With tighter budgets, I see all of our clients searching for ways to reduce training and travel costs without cutting training programs. They want to continue to develop their employees and managers without sacrificing the quality of instruction. My advice? Let Cynthia Clay and this book help you achieve those aims.

As for PACE, we will continue to be one of Cynthia's webinar students, and wherever possible, one of her disciples. We can provide personal testimony for the notion that "you, too, can deliver great webinars."

Jeanne Knutzen,
Founder/CEO, PACE Staffing Network

Preface to the Revised Edition

Since the first edition of this book was published nine years ago, both technology and the training industry have experienced changes. When I wrote the book initially, most training was still classroom-based, but organizations were waking up to the realization that virtual learning could reduce expenses and employee downtime. However, most organizations had not yet experienced the improvements in learner engagement and retention that the technology could bring. I'm proud to say that the original edition of this book helped to change that mindset. Today, it's often taken for granted that training events will be offered virtually.

When I wrote the original edition, there were only a few platforms that supported the kind of interactivity we were looking for in creating immersive learning experiences. Our favorite was, and still is, Adobe Connect (then known as Adobe Connect Pro). As virtual meetings became more common in the workplace, we've seen rising popularity of simpler platforms such as Zoom. These platforms are affordable and easy to learn, and work great for meetings, but often lack more sophisticated design features that improve the learning experience, such as layouts in Adobe Connect.

Also, over the decade many users have gotten accustomed to being on-camera. But this remains an area where I hope to see more progress. Your on-camera presence is so much more engaging than voice-only.

A number of significant changes over the last decade come out of mushrooming content platforms such as YouTube and the ubiquity of mobile devices. Curious learners don't have to wait for classroom training to be offered – they can Google pretty much whatever they want to know, whenever they want. This doesn't diminish the importance of formal training programs, but it does change the participants' expectations of the process, and how they choose to participate in that process.

One outcome of these changes is the rise of microlearning, or chunking learning content into small units that are readily available. Meanwhile, the fact that learners are no longer meeting physically in a room has focused attention on the human need for connection being addressed in the virtual, leading to platforms dedicated to social learning.

In addition to these revolutions, over the last ten years, I developed an interest in the neuroscience behind learning. This was sparked by discovering that both of my daughters were dealing with learning disabilities (ADD and dyslexia). As I watched them struggle to achieve in the traditional school classroom, I wanted to understand more about how their brains were wired. One of my favorite resources is *Brain Rules*, written by a neuroscientist, John Medina, who is a professor in Seattle where our company is based. My investigations also took me to Morningside Academy, founded by Dr. Kent Johnson, who developed an approach called generative learning, to help intelligent kids with learning disabilities thrive in educational environments. Morningside Academy's work with students includes rigorous practice of basic, foundational skills to anchor them as knowledge they can access quickly; daily goal setting; consistent feedback from teachers; and personal tracking of their improvement in test scores.

Our older daughter, Grayce, is a testimony to the generative learning approach. When she entered Morningside Academy at age nine, she was barely able to read, write, and do math calculations. It was painful to hear her cry at night over her homework, "I'm too stupid to do this!" I knew she was intelligent and I was flummoxed at how best to help her discover that for herself. Thanks to Morningside, she is now a hardworking student in college where she just achieved a 3.9 grade point average in her first semester as a freshman. Our younger daughter, Jessica, is also doing very well in high school and aspires to be a neurosurgeon. Whatever they decide to do for their future careers, they've both become well-acquainted with their unique needs and capabilities as learners.

My personal interest in this topic led me to apply brain-based learning principles to our work in the virtual classroom. I realized that these principles can be used to ensure long-term retention and application of the topic. My musings led to me to write and speak about 12 brain-based learning principles, six for engagement and six for retention. I've added Chapter Ten to this book to cover this important topic.

Acknowledgments

There are many people I want to thank—for their mentoring, encouragement, support and just plain hard work over the past few years. I will name just a few of them here.

My friend and colleague, Paul Petrucci, took my web workshops, virtual courses, bits of writing, assorted presentations, and odd snippets of interesting observations, and somehow molded them into a coherent book that makes me proud.

My smart husband, Leo Brodie, brought his editorial eye to the first drafts, challenging, prodding, questioning, and correcting. It is a much better book as a result.

Jeanne Knutzen, a dear friend, business owner, and mentor, first suggested that I might be on to something with this webinar business and encouraged me to pursue it. Bev Kaye, a marvelous role model for anyone who wants to write and speak to others, inspired me to want to write my own book and has been a great cheerleader for my success.

Marianne Cherry, my childhood friend and also a former director of client development at NetSpeed Learning Solutions, introduced me to Adobe Connect, seeing the possibilities of this web conference platform before I did. Tim Jones, formerly our VP of sales/marketing, embarked hand-in-hand with me down the web conference path, collaborating with me as a host, as we figured out how to deliver engaging webinars together. Elaine Smith, operations manager, led the book through its final stages of production.

The speaker selection folks for the Association for Talent Development (ATD) International Conference first invited me to speak about virtual facilitation at the annual conference in 2008. That session helped me realize just how many people were hungry for creative, interactive web training.

The first participants of the VFTC course took a test drive, liked what they saw and gave me very helpful feedback to improve the experience for those who came after them. They include: Fred Allemann, Mimi Banta, Frumi Barr, Robert Black, Brant Blumstein, Carole Bourque, Louise Carnachan, Theresa Chambers, Marianne Cherry, Anne Davis, Andrew Feldman, Lynn Gaertner-Johnston, Jackie Galleano, Jen Gregg, Mary Harter, Lorraine Howell, Wendy Jackson, Bernice Johnston, Sue Johnston, CJ Keyes, Nancy Kohutek, Michelle Kunz, Patricia Luzi, Deanna Maio, Jeanette Nyden, Karen Pacent, Tonya Perpich, Stan Pietrzak, Dave Ritter, Vernon Roberts, Christine Sandulli, Lynda Silsbee, Janet Spadola, Sandra Starr, Katharine Wismer, Veronica Zabala, and Amy Zinman.

And last, but never least in my heart, my two daughters, Grayce and Jessica, who have met the challenges of ADD and dyslexia to succeed in high school and college. I am continually inspired by their achievements.

Cynthia Clay
March 2019

Introduction

After 25 years in the training and development industry, I thought I was a pretty effective classroom facilitator. So when my clients began to request that I deliver webinars, I thought, "how hard can this be?" I was curious to see whether web conferencing tools could be used to replicate or replace the classroom experience. But I was little prepared for how much I had to learn. Perhaps you've picked up a copy of this book because you feel the same way.

My early efforts were mediocre at best. I experimented, tinkered, requested feedback, and absorbed blistering comments from disgruntled and honest webinar participants. Let me share a couple of the most memorable observations that had an impact on my thinking.

"I'm so frustrated. You droned on and on at the beginning of the webinar. So I started to multitask. You got into the meat halfway through but I missed the part I really wanted to hear. Bummer."

My first response to that comment was to scratch my head and wonder aloud, "Is your multitasking my fault?"

Well, what if your multitasking really is my fault? If I accepted that responsibility, what would I do differently to keep you engaged?

Here's another one:

"Oh my God. Not another boring, bulleted PowerPoint slide. Point by point, slide by slide. I'm s-l-o-w-l-y dying of boredom....help me.....aaarrrrggghh."

I must say I laughed out loud at that graphic description of disengagement and vowed never to get that response again. (Thank you, oh bored participant, wherever you are.)

I've tried and abandoned multiple web conference platforms because of odd technical limitations, clunky presenter interfaces, and confusing participant experiences. Perhaps, you too, have been frustrated by the web conference jungle out there. We finally landed on Adobe Connect, a platform that allows for creative, interactive design and delivery.

Along the way, I've been blessed to collaborate with client partners like the United States Tennis Association, XO Communications, the US Government Accountability Office, Grainger, Genentech/Roche, the Nature Conservancy, and Cardinal Health, who helped us get better at developing and delivering great webinars.

I'd like to be that partner for you. I've written this book to help you learn what my clients and I have learned about what works and what doesn't in the world of virtual delivery. In this updated version, I've added a new chapter exploring 12 brain-based learning principles that are the foundation of engaging, lasting virtual learning. I want to help you create memorable, collaborative virtual training experiences for your audience. And I hope you'll come to love the possibilities of virtual delivery, not just to reduce your training and travel costs, but to create truly engaging web workshop experiences.

Cynthia Clay
NetSpeed Learning Solutions
Revised and Updated in 2019

chapter one

virtual facilitation gone wrong

Technology has allowed the business world to expand geographically, but economic challenges have caused our travel budgets to compact. This presents challenges for educators. They can't always get participants in the same room, or even the same country. Webinars hold promise as a replacement for the physical classroom. Simply package up your product. Eliminate the classroom interaction since that's not possible in a virtual setting. Present your material, intact, over the Internet. Problem solved, right?

Wrong. Effective learning transfer requires participant engagement and collaboration. No matter how modulated, a disembodied voice presenting one-way to participants on PCs is a recipe for turn-off. A tedious, non-interactive presentation breeds boredom in the participants. Associating the webinar environment with boredom diminishes expectations. Diminished expectations cause your students to come into the training session ready to give it only half their attention.

It's a vicious cycle. My purpose behind writing this book is to help you break that cycle. There is a science and an art to facilitation in the virtual classroom. This book will explain those scientific and artistic techniques. And it will begin by examining why webinars have a bad reputation.

(Don't be concerned about the somewhat gloomy start to our study of webinars. We'll get it out of our system in this chapter, then go on to explore tools, tips and strategies for building powerful web workshops that are captivating, informative and fun.)

In the article *What Stinks About Webinars*[i], co-author Allison Rossett encouraged her students at the time – co-authors Antonia Chan and Colleen Cunningham – to feed her their positive and negative impressions of webinars that they attended. The majority of impressions turned out to be on the negative side.

We carry out the same exercise at NetSpeed Learning Solutions. Before the first session of our *Virtual Facilitator Trainer Certification*[ii] (VFTC) course, I ask the class participants to sit in on as many webinars as they can. It's very helpful to see things from the side of your audience. The pain you may suffer as a participant is a great motivator to raise the bar of quality on your own web training sessions. And it helps you understand how your audience will perceive your webinar: why they might dread showing up, and why they might be tempted to multi-task.

NetSpeed Note
Our mission at NetSpeed Learning Solutions is to develop better leaders and more engaged employees through both face-to-face and virtual learning. Our Virtual Facilitator Trainer Certification (VFTC) course is intended for anyone who wants to design and deliver engaging, highly interactive web conference training. It's a four-week intensive workshop, delivered through synchronous, facilitated webinars and self-paced, asynchronous online content.

The student assessments reveal certain common mistakes of webinar presentation. See how many of these mistakes you can recognize from your own experience on either side of the podium.

Putting up with poor preparation and content

> *If you take an already bad training presentation, poor content and organization, throw in terrible and boring slides, add a weak, talking-head facilitator, what would you get? Answer – the one hour webinar I just watched. Honestly, I could not get through all of it – too painful.*
>
> ~ *Jackie*

Poor preparation and content are a devastating combination in any training setting, whether traditional or virtual. Ho-hum slide decks are more the rule than the exception. In many companies, the terms PowerPoint presentation and snoozer are synonymous.

The reasons for this are well catalogued. Too often slides are crammed with bullet points holding too much information (and too few graphics with emotional or visual impact), and presenters who vocalize from them verbatim, often in a monotone. When slides are presented in this way, you may as well email the slide deck – people can read more quickly than they can hear.

As learners, we empathize with the struggle it requires to sit through a poor presentation. As instructors, we must learn to pinpoint the reasons for it. So, applying the basics of good slide design is your first, and perhaps easiest, step.

NetSpeed Note

Too often the sequence of slides in the deck, their volume, and the use of graphics, graphs, and lettering break every rule in the book.

We'll present some tips for design and delivery in Chapter 5.

Besides telling you that sloppy slides are a bad thing, here's another duh: poor delivery of those sloppy slides adds insult to injury. The consequence of not fixing basic PowerPoint design and delivery mistakes is that your audience will zone out.

Here are some more basic presentation mistakes that you don't want to repeat, chronicled by my webinar students.

I would like to have seen some learning objectives at the beginning and perhaps a review of the learning objectives at the end. I generally got the most important intent but wasn't sure what other objectives the instructor had in mind. ~ Fred

It was hard to follow the presenter, as he went through some very complex steps very quickly. He didn't pause to explain terms that some of the participants might not have understood. He also tried to cover too much in the session. ~ Mary

Yearning for something to watch, I started focusing on the questions coming in the chat box, rather than the talk. Ooops! ~Sue

A good portion of the beginning of the session was used to talk about the three presenters' job titles and levels of experience. Although I appreciate knowing a bit of relevant information about the presenters, I started to lose interest as the minutes ticked away. Remind me, why am I here again...? ~ Carole

The facilitator would display a slide with about 4-5 points. Then she would talk about the points for what seemed like several minutes without necessarily referring to them specifically or using a pointer. As a visual rather than an auditory learner, I found her approach challenging. ~ Lynn

There was nothing for the kinesthetic learner. I really struggled to stay engaged. ~ Sue

I was reminded of the importance of keeping promises, when the Q&A period was significantly shorter than the presenter had promised. I am sure that had I been in a traditional classroom, I would have seen frustration on faces when there was little time for their questions. ~ Lynn

As you'd expect, the one-to-one conversion of flawed slide design and inferior presentation techniques from a classroom setting to a virtual one doesn't magically improve or eliminate any of the above issues. As a matter of fact, just rehashing a less than superior speech in a webinar format makes the experience worse. The participant is not held as strongly to the social pressure of paying attention, or even of acting like they are paying attention.

This is probably the major consequence of poor presentation in the virtual arena: participants have more ways to tune out. Unlike the physical classroom setting, they are sitting in front of their primary tool for getting work done: their computer. The temptation to multi-task takes over, and who can blame them? They may rightfully consider that catching up on work adds more value than listening to text they can easily read later. Of course, since we are all very busy, that later reading never takes place.

As one of my students wrote in her assessment:

I think this is why people end up multitasking. I'm writing the critique now while she talks because I'm not missing anything on the visuals. If she would only engage me and make it hard for me NOT to participate, I would not be able to do this.

I've got one ear open for something that catches my attention, but meanwhile, I have seen how this works, and it doesn't require me to be fully present. ~ Michelle

If your presentation suffers from poor slide design and poor delivery in the physical classroom, you've got a problem. If you merely transfer that presentation to the web, your problems are compounded.

People don't want to fall asleep. They want to be engaged. Which brings us to the next observation gleaned from an analysis of webinars.

Looking for interaction and collaboration

I was appalled at the lack of interactive content. The first few minutes gave me great hope: A host gave clear instructions on how to use all kinds of participant tools and then introduced the presenter. But once she got on, it was a one-woman show. At one point the host even tried to interject an observation, and she actually plowed right over him and continued talking. I had felt RELIEF that another voice had introduced a discussion point, but then felt horror when she ignored him and doggedly went on with her agenda. ~ Michelle

Educators can do a creditable job of trying to be conversational, using an animated and well-paced delivery. But without real involvement of your learners, you're disregarding one of the critical findings of adult learning theory: that your audience wants and needs to bring their experience to the table.

Interaction between the facilitator and participants promotes interest and engagement; collaboration among the participants promotes sharing and provides the basis for learning activities. The days of having students passively sitting and listening should be long gone.

All good web platform tools contain interaction tools that allow interaction and collaboration. They all make use of conference calling that allow all parties to speak. They all provide streaming media so you can see the presenter. So what gives? Why aren't these tools used effectively, so that trainers and facilitators can put good learning practices into place as they do in the physical classroom?

Part of the answer may be that it's simply easier not to. With webinars there is, without a doubt, an additional challenge of managing the technology while managing the presentation. But there may be another answer. One that has to do with a lack of belief.

In my facilitator training classes, I start out by asking my learners to explore their own attitude about web training. Like some of them, your experience of learning in cyberspace may be scant. The overriding negative belief is that virtual classroom environments don't encourage participant involvement because they can't. Since webinars provide one-way communication, webinar instruction contains challenges that can't be overcome.

Here is some of the thinking that lies behind negative notions about webinars when it comes to collaboration and interaction.

- It's difficult to facilitate virtually because I can't see the learners' body language or facial expressions.
- I want learners to engage with each other, not just with me. But that's not technically possible in a webinar.
- I think web conferencing is a one-way communication tool. It does little to foster a sense of community.

- Most learners prefer to attend virtual classes because they can multi-task while the facilitator lectures. This is actually an advantage of this type of learning.
- Virtual training is a more efficient use of time because you can give learners the information quickly and they can get back to work.

What these negative beliefs have in common is the assumption that when addressing interaction and collaboration, web instruction is a brave new world; you can't use interactive techniques, tools, and exercises in the virtual environment.

Of course, I don't believe it, or I wouldn't have written this book. But let's keep the conversation going about why most webinars have a bad name, since knowing what *not* to do can be very useful.

Getting instructional materials and class size wrong

In the previous section, we discussed the widespread belief that webinar environments don't encourage participant involvement because they simply can't. You won't be surprised to hear that facilitators hold misconceptions about instructional materials and class size related to webinars. Here are a few.

- Instructional materials are used differently for web workshops.
- Learners, in a virtual setting, do not need handouts.
- Learners should receive a copy of the slide deck prior to the session so they can follow along.

I'm going to suggest that these negative notions about classroom materials, as well as the misunderstanding of the potential for interaction and collaboration discussed previously, are in large part based on this single underlying belief:

- Webinar technology can reach many more participants, so you can educate hundreds just as effectively as twenty.

In other words, because web conferencing technology allows us to speak to hundreds of listeners in a one-way monologue, then it's a good idea to do so.

If you accept this as the de facto webinar model, then pessimistic opinions about webinars will definitely prove to be correct! If you have a hundred listeners on the call, you're constrained on all sides – by the technology, by time management, and by people management.

You simply don't have time to solicit audience contributions, or even their feedback. You can't open up phone lines or you'll cause havoc. (Not to mention that there will be no phone lines to open, since you're probably using VoIP technology for an audience that large.)

NetSpeed Note
VoIP stands for Voice over Internet Protocol. It provides audio that is heard over the learners' computer speakers. It's a very cost-effective for large audiences, but, often, not as practical as using a standard teleconferencing solution for small audiences. *You'll find a glossary of terms in the back of the book.*

You can't manage logistics of breaking people into groups. You also can't have them use the whiteboard. Collaboration among your audience is a non-starter.

Speaking of logistics, you can't hope to have everyone in the audience receive handouts and be pointing to the same page. You might as well give up on handouts.

You may also lose an essential tool for holding attention: streaming video. With large audiences, using both VoIP and streaming video may cause a rough audio transmission that is out of sync with the video. So streaming video of your animated, excited face is replaced by a photograph of your animated, excited face. Not the same thing at all.

Having large numbers of webinar attendees is fine for marketing webinars, one-time information-packed presentations. For the standard learning session, large numbers of participants are as problematic in virtual training environments as they are in the classroom. We'll talk more about that later.

Encountering trouble with technology

To err is human, but to really foul things up requires a computer.

~Farmer's Almanac, 1978

The traditional classroom may leverage technology, but the virtual classroom depends on it. A technology hiccup in the virtual world promises an unpleasant learning environment at the least; at the worst it can shut down the class completely.

Stuff happens. It happens to different components (like software, hardware, and network) at different locations along the line, from facilitator to web platform company to participant. Moreover, the symptom of the problem being suffered by party A may have a root cause that has party B to blame. Yes, technology tribulations can be one big happy party.

Bandwidth issues are the most ubiquitous. Bandwidth refers to the amount of data that can be passed over the network in a given amount of time. It's synonymous with "connection speed." Do you remember having to connect to the Internet over your phone line? It was slow! Thankfully, today our workplaces provide sufficient bandwidth for streaming video and using VoIP. But because your participants have access to fast connection speeds, it doesn't mean that they've set their software properly to experience fast connection speeds. When they do it incorrectly, their problem becomes your problem.

It's starting to sound like you have to add "Technical Guru" to your resume. The good news is you're not expected to singlehandedly solve all issues. The bad news is that even if you're not enamored with bits and bytes, you should at least be on speaking terms with them. That translates as being able to quickly identify symptoms and causes, knowing whether they can be resolved, and knowing how they can be resolved.

And unfortunately, you do have to (and will) experience the rite of passage known as glitch initiation. After that, you have the right to tell your war stories to anyone who will listen.

Okay, I know you're dying to hear mine, so here goes.

NetSpeed Note
Actually, there is a way to escape glitch initiation. Facilitators often have a co-anchor, or host, who introduces them and has the responsibility of all things technological. We'll discuss that option more in Chapter 8.

At that time, my company had a 100-seat license for our web platform tool, Adobe Connect. Over a hundred eager learners signed up for a marketing webinar. I was unavoidably running late (never mind why), so by the time I arrived, my panicked associate had opened the room for participants to join. The session quickly reached the 100 member capacity. We were delighted with the attendance. But there was one problem. I couldn't log in to my own webinar.

Some hardware and software problems haven't been my fault, I swear. Another time the conference call line for our large public webinar with thousands of participants was joined by a different meeting with 20 people. Somehow their group had been assigned the same teleconference number as ours. Their group leader had to send out another email to correct the problem. Unfortunately, for the remainder of our session we were sporadically joined by new visitors wondering why we weren't teaching *How to Make Millions in Credit Default Swaps!*

I've had PowerPoint slides freeze so that they refused to advance. I've had the Internet go down in the whole region, abruptly ending our web workshop. On one occasion two of the three servers on the web platform failed, which made the performance really slo-o-ow for everyone considerably.

Other times, I'll admit, the fault was my own. Once I muted my phone line to cough and take a drink of water while participants worked on an activity. When I came back on camera, I started debriefing their chat messages. I was happily talking to myself for at least a minute before I received a chat message from a kind (and bored) participant telling me I was still muted. Another time I didn't replace my phone's headset on its recharging cradle the night before, so it was out of juice when I ran the web workshop. I had to hold the phone handset to my mouth the entire time, a constant reminder to those watching the video of my incompetence. (I won't even mention how hard it is to manage a webinar with one hand.)

Urban legends abound. One facilitator I know used web platform software that closes a meeting after the facilitator signs off. A few minutes into the session, my friend was somehow dumped. They couldn't get in again. Webinar over.

Web servers can go down (hey, it happens). There are latency issues. Everyone – the entire cadre of participants and facilitators – can get kicked off of an event. The screen resolution is different for different users, so that some can't see parts of the screen that others can. Some platforms make the participants download software, so the number of potential mishaps extends to their firewall, spam filters, popup blockers. Not to mention that they have trouble loading the software itself.

Isn't this fun? And I haven't even mentioned the web platform feature that reduces even large, grown facilitator men to tears: breakout rooms.

Breakout rooms offer the promise of easily placing participants into small groups. In their cones of isolation the teams can then collaborate amongst themselves, while the facilitator pops in and out of each room, offering wisdom as needed.

What the fine print in the software neglects to mention is that there are numerous ways, some of them quite creative, that breakout rooms can break down. In some cases your students, whom you have come to know and love, are lost in cyberspace, never to be heard from again (at least in that session).

And while you're desperately seeking your students, guess what happens if you're recording your session for future playback? Yes, that's right. The recorder stays right with you in the room, capturing for posterity your frantic calls, curses, pleas and appeals to a greater power.

Yup, I've been there.

Offering hope

Whooo! We've examined poorly done webinars and have diagnosed their weaknesses in a number of areas, including poor presentation, interaction, collaboration, instructional materials, class size and technology. But don't worry, there is hope.

In the next chapter, we'll touch on ways to improve this sad state of affairs. We'll step through a simulated webinar that gives you a taste for how to immediately engage students in interaction and keep them from tuning out. You may begin to see how beliefs about the inadequacy of webinars are less "correct" than they are self-fulfilling prophecies.

In following chapters, we'll delve more deeply into how to deliver great webinars. Chapter 3 addresses how to get to know your audience and your objectives. I'll model how to promote effective interaction and collaboration in Chapter 4. In Chapter 5 I'll present tips and techniques for using PowerPoint slide presentations on the web. Within that conversation, we'll touch on when not to follow standard PowerPoint practices in a virtual setting. We'll go over how to re-purpose your traditional classroom exercises to the virtual classroom in Chapter 6. In Chapter 7, we'll discuss adult learning theory and how learning transfer takes place in the virtual setting. In Chapter 8, we'll offer some pointers on how to prepare yourself for addressing the inevitable technology snafus. Chapter 9 illustrates how to prepare and present a dry run of your web training.

As we step through these tips and techniques, I'll continue to bring in student comments and present best practices gleaned from the community of virtual facilitators.

NetSpeed Note

Speaking of the community of virtual facilitators: there are several national associations that can help you improve your skills in both the virtual and traditional learning environments.

*American Society for Training & Development (*http://www.astd.org/)

*National Speakers Association (*http://www.nsaspeaker.org/)

*United States Distance Learning Association (*www.usdla.org*)*

*The International E-Learning Association (*www.ielassoc.org*)*

*The Masie Center (Learning Lab & Think Tank) (*www.masie.com*)*

*Society for Applied Learning Technology (*www.salt.org*)*

The eLearning Guild (www.elearningguild.com)

Homework

We learn better by doing than by reading. At the end of each chapter, I'll offer some options for self-study and practice. By the end of this book, you'll have developed the basic skills necessary to deliver interactive online training using any web conferencing platform. You'll be able to:

- Design effective virtual classroom exercises using web conference interaction tools
- Engage learner attention and participation in online learning
- Repurpose traditional classroom exercises for collaborative online learning
- Describe best practices for online facilitation
- Develop a comprehensive plan, including a simple facilitator guide, PowerPoint presentation, and participant materials for a 20-minute online learning session.
- Increase learning transfer after the facilitated session

Assignment 1-1: The Virtual Facilitator Self-Assessment

Read and fill out the Virtual Facilitator Self-Assessment in Appendix A. After completing it; write a one-page response on the advantages and disadvantages of synchronous (i.e. two-way, real-time) web conference training.

Assignment 1-2: Ebook: "Bring Your Mojo to Virtual Learning"

Read this short ebook to explore how to bring your best to webinar training delivery. You'll find it at

https://netspeedlearning.com/contact/?p=ebook&r=mojo

---

[i] Rossett, Allison. "What Stinks About Webinars." Chief Learning Officer August, 2008: http://www.clomedia.com/features/2008/August/2317/index.php

[ii] NetSpeed Learning Solution's *Virtual Facilitator Trainer Certification* course is designed for experienced classroom trainers who want to become experts at designing and delivering engaging, highly interactive web conference training. Read more about the class and the certification here:

https://netspeedlearning.com/virtual-learning/facilitation/

chapter two

getting it right

The word webinar means "web seminar." In a seminar you simply sit and watch someone present. We learned from the previous chapter that this is anathema when trying to engage participants in a learning environment, be it virtual or traditional.

In our attempt to change whatever negative notions you might hold about webinars, let's start with a name change. Let's refer to an interactive, scheduled eLearning experience that occurs in real time with an instructor or facilitator as a "web workshop" instead of a webinar. This term more correctly captures the sense that you will be *doing* something, instead of passively listening.

Presenting a web workshop example

As mentioned in the previous chapter, my company holds web workshops for trainers seeking certification in delivering virtual learning sessions. In these workshops, we facilitate facilitators. That is, we present content and techniques for virtual facilitation, while simultaneously modeling the content and techniques. In this way, participants can learn through immersion.

In this book, I'll be using our *Virtual Facilitator* web workshop as an example, not just for its content but for the way it models the techniques and best practices for effective web workshop delivery.

I've taken a creative approach here to not only demonstrate these techniques but more importantly, show how they impact the participant. I've created a fictional character, Gerri Jordan, a Customer Service trainer for a Fortune 1000 corporation. Gerri is attending my workshop. I'll describe the session through Gerri's eyes, ears, inexperience, and eagerness to learn.

So yes, this means I'll be writing about a fictional character experiencing my own actual workshop. It may seem odd that I'm writing about myself in the third person (and even odder that I'm including remarks about what a wonderful presenter I am!). But as facilitators, we should always be thinking about how our audience will experience our presentation. What follows is a depiction of the kind of reaction I hope to achieve. As you develop your own web workshops, I encourage you to imagine eager, engaged and possibly skeptical participants on the other side of the screen, as I have done here.

> Gerri sits in front of her PC with her headphones on. She has been encouraged – no, let's say forced – to add virtual sessions to her duties. She's not a techie and she is a bit fearful that this change of venue will make her less effective in the eyes of her management, not to mention her students. This apprehension has been reinforced by her attendance in web seminars (the boring, badly done type).
>
> Gerri clicks the link in the email message that introduces the web workshop, and dials in to the conference call. A welcome slide for the day's class is displayed, but the presenter hasn't yet arrived.
>
> Gerri has three Word documents open in other windows. She is working on one of them, the upcoming class on Customer Server that she will be giving, when the facilitator's voice is heard.
>
> "Hi, everyone. Welcome and thanks for attending!"

I'll just keep working on this document until I have to look at the slide. I hope she goes over how to mute the phone since it's a little noisy here.

She's surprised to hear multiple responses from participants. "Hello." "Hi." "Good morning."

This is a first. People are actually talking.

Intrigued, she changes to the class window. A streaming video shows the image of a woman, her smile beaming brightly. Gerri watches her animated expressions.

> "I'm Cynthia Clay, President and CEO of NetSpeed Learning, and I'd like to welcome you to the first web workshop of the *Virtual Facilitator Trainer Certification* course. Today we're going to create an engaging virtual learning session that leads to learning transfer.
>
> "Before we get to that, let's talk about some housekeeping."

A new slide with the title "Housekeeping" is displayed. A green arrow appears next to the first bullet.

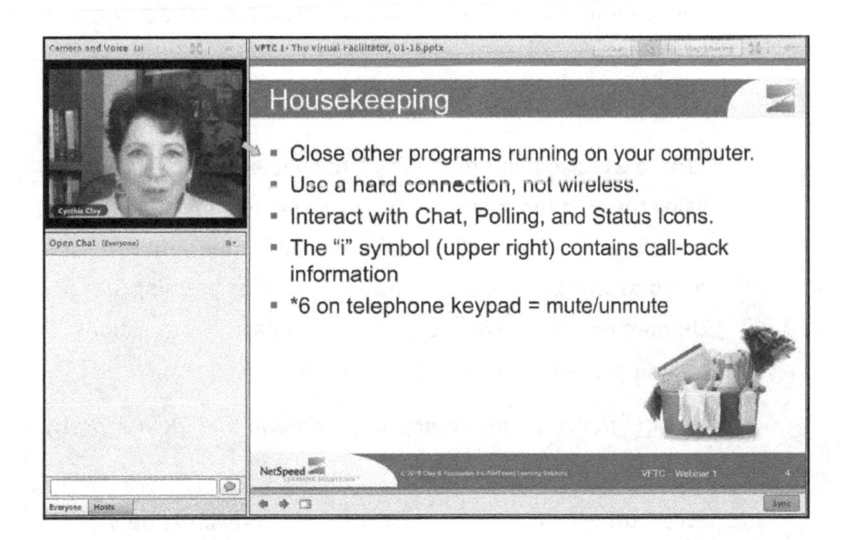

Oh, great. Bullet point parade. I wonder how long this will take.

"These housekeeping rules will make the experience more positive not only for you as the eventual presenter, but also for the participants.

"For best results today, I'd like you to close out other programs that might be running on your desktop. Web conference platforms use lots of bandwidth, and you may find that having other programs open interferes with the performance of Adobe Connect."

Hmm. I was hoping to get my project done during this webinar. I won't be able to do that if I close the program. I wonder if we can ask questions during the session. Well, I don't want to bring that up now because that will slow everyone one down and I could look foolish. We're off to a bad start.

"During this session if you have questions, send them right to me or our host, and we'll help you. Just use the Open Chat box below my video. This is a great way for us to interact with you without distracting everybody. If you send us a private message, only we can see it."

Oh. Okay.

"There will also be opportunities to chat with each other, using the chat function, and we'll use polling and status icons to communicate as well, which we'll talk about. And you might as well go ahead and mute yourself by pressing Star 6. Unmute by pressing Star 6 again. If you have any questions about this, just use the Open Chat box."

Wow, that green arrow is hopping through the points pretty quickly. I like this pace.

"We'll provide a brief survey at the end of this session. At that time we'll also send a slide deck out to you."

Why at the end? I usually give slide deck handouts at the beginning of my class.

A slide with the title "Learning Goals" is displayed.

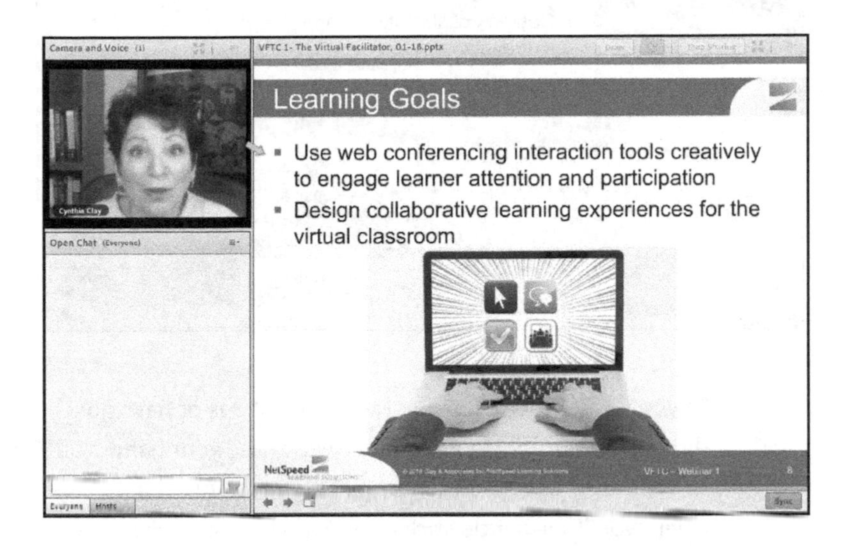

> "Now that we have the housekeeping out of the way, I want to welcome you again. Some of you I've met before, and some I haven't. I'm going to pull this Attendee List to the center."

An Attendee List appears above the slide.

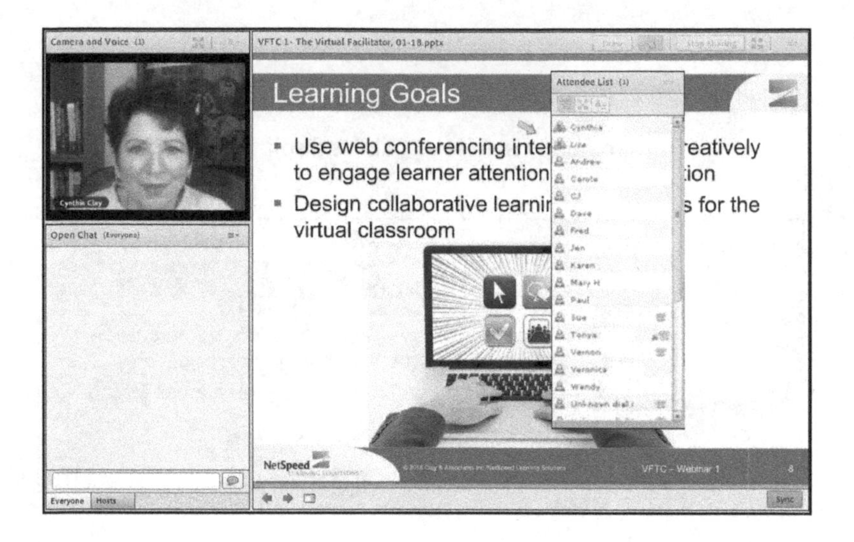

"If you've attended one of our web workshops or have gone through certification, I'd like you to just raise your hand. You can find the raise-hand icon by looking above my video, where you'll see a little stick figure with his hand raised."

Well, I haven't attended before, so I'll sit tight. Hmmm, look at those hands being raised. John, Jennifer, Lydia. I wonder where they're from.

The presenter continues speaking as hands go up, waiting for all respondents.

"So, it looks like there are a few people new to us. Today and in the next three sessions we're going to look at web conferencing and focus on how we can make this forum highly interactive and engaging to people. Throughout we'll attempt to model some of the tools. Our web conferencing tool, which today is Adobe Connect, has many features that allow you to get pretty creative for learning design.

"Many of you are going to be taking programs that already exist in the classroom and re-purposing them for virtual learning. We'll focus on that in our third class."

Yes, that's me!

> "Let me find out how much experience you've had with web conference delivery. One of the things I like to do is get you engaged in the tools. We don't talk much about how to use them at first, we just jump right in and use them."

A poll appears, sporting four options. The number of responses jumps as participants answer the poll. Cynthia continues talking throughout.

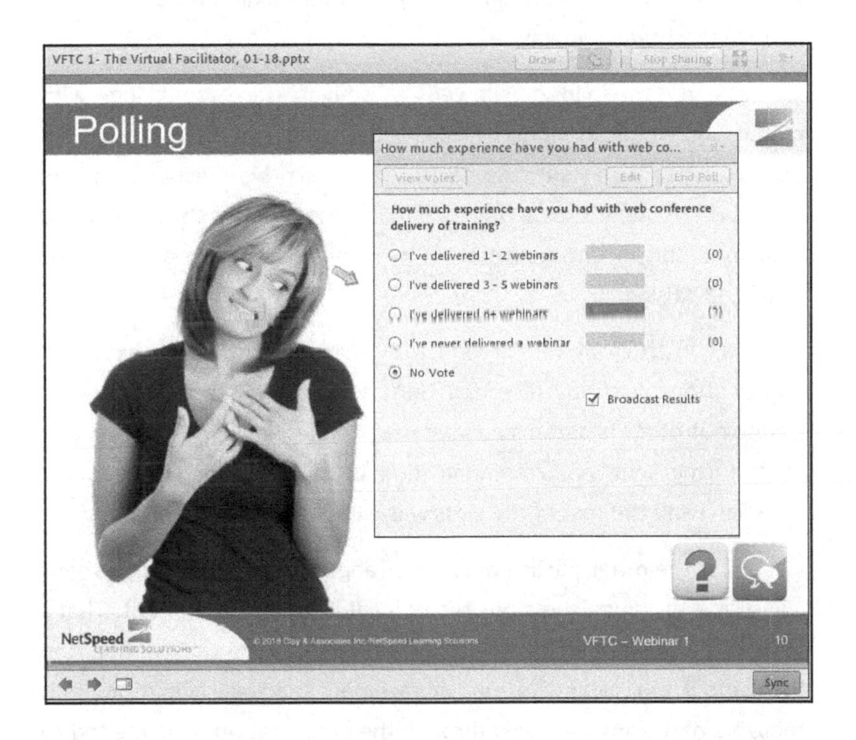

Let's see. I've attended four webinars, each a miserable experience. Wow, this one is so different! She seems to have captured my attention. How the heck did she do that?

Let's pause here, at the six-minute mark of the web workshop. As we discussed already, one of the greatest challenge of workshops is to reproduce the warm environment of a physical classroom session that will engage the participants and facilitate learning transfer. You need to do this right off the bat.

Capturing attention

With a little more reflection, Gerri will realize how her attention was captured. For starters, she anticipated a boring, droning tutor talking out their PowerPoint bullet points, and was set to listen to the presentation with half an ear, to begin multitasking, and even to get up to take a break midway.

Instead, the video showed a smiling face speaking with modulation who seemed to be looking her in the eye. Even though the instructor had no visual cues to see how Gerri was responding, she immediately invited two-way communication, both through audio responses and in interactive tools of chat, polling, and status icons (the hand-raising).

The instructor reinforced the notion that she expected participation by subtly directing Gerri to disengage her other work ("close out other programs and set your connection speed"). Then, by experiencing two-way communication instead of one-way, rapport with Gerri and the rest of the class was established and maintained.

So far the other participants are strangers to Gerri, which perhaps is just as well, since she's not ready for full-scale involvement. That's similar to what happens in a physical setting. But like that setting, the foundation has been set for engagement and participation, and thoughts of drifting passively through the presentation have melted to the background.

Keeping the class size small

Gerri has seen the attendee list, and knows that she's one of a manageable group of people.

In the previous chapter we discussed the misconception that because you *can* get hundreds of participants on a webinar using VoIP, it's a good idea to do so. Successful learning transfer requires interaction and collaboration. You can't experience that as keenly with large numbers.

In the traditional classroom, the best class size is from fifteen to twenty. Keep this same range for virtual classes.

NetSpeed Note
Having said this, chat, polling, and the raise hand tools can be applied to marketing webinars with larger audiences. Check out Appendix G for some hints on how to do this.

Smiling for the camera

How do you replicate the warm web environment that Gerri and her co-learners experienced? Let's start the discussion by addressing a single factor: streaming video. Before the skill becomes second nature, delivering your presentation or training in front of a web camera without visual feedback may feel like speaking inside a closet. Besides the discomfort of being on camera for an hour or two, some facilitators fear it will distract their learners.

Don't let that discomfort and fear dissuade you from using streaming video. It actually adds another visual element that captures attention and builds connection. It also helps to remember that you are speaking to real people with real needs—talk to them. That's easier if you know something about them, as we'll touch on in the next chapter.

Even if you can't see them, your audience is processing a lot of information about you through your smiles, gestures, and facial expressions. But for greatest impact, you need to maintain the "illusion" of eye contact. Remember to look in the camera lens. It's critical to maintain awareness of the camera. If your eyes tend to wander to the screen, place a sticky note there to force you back to the camera.

Using interactive tools effectively

I mentioned two-way communication. Most web conferencing software programs feature an abundance of interactive tools that provide two-way communication. Yet some of them are utilized only lightly by presenters, or not at all. Why is that? As suggested in the previous chapter, it's partially due to a belief that physical classroom techniques can't be repurposed for the Internet. This notion may come from not having seen it done effectively. After all, it's one thing to be aware of a tool, and another to know its strengths and when to use it.

NetSpeed Note
Speaking of interaction tools, you'll find a list of them in Appendix B. For each tool, there is information on its uses and benefits.

Let's take the raise hand icon as an example. This tool is intended for situations where you need a "show of hands." In the web workshop example, it was utilized early in the session to see how many people had attended previous sessions. Responses in this situation can provide several benefits.

For the facilitator

- In recognizing individuals from earlier classes, the facilitator might tailor content, questions or their style of teaching differently.

- The tool gauges the number of active participants.
- Knowing the percentage of participants who were familiar with the housekeeping rules and various tools might affect the pace of the session.
- Forcing a response encourages participants to focus on the class.

For the participants

- Being able to view all responses allows a participant to see where they stand in relation to their classmates.
- Participants know the size the class. Knowing that the class is small encourages participation.
- Participants see the names of the people that they may get to know better during the session.

That's a lot of knowledge and behavior change from a simple teaching aid, is it not? As this brief example shows, there are multiple uses for even the humblest of tools in the virtual toolbox. We'll spend more time on each of the interaction tools in Chapter 4.

Perhaps you're beginning to see that interaction tools are a key to emulating the physical classroom environment, and their use is at once subtler and deeper than they might appear at first glance. The takeaway is straight from the learning goal stated at the beginning of this chapter: the understanding and intelligent usage of interaction tools is necessary to engage learner attention and participation in the virtual environment.

And if you can engage learners, you've already won half the battle. The next chapter will discuss how to get to know those learners better, and how to use that knowledge to impact the design of your web workshop.

Homework

Assignment 2-1: Familiarity with interaction tools

Read about the uses and benefits of interaction tools in Appendix B. You'll revisit this list in Chapter 4.

chapter three

know your audience / know your objectives

The planning of any project requires setting clear, measurable objectives. If you don't have a measure for success, you can't tell when you've reached it. The same goes for web workshops.

Whether you're designing a single slide or a group of slides, a single session or an entire workshop, you must design to measurable objectives. We tend to pound out PowerPoint slides and present them without asking ourselves probing questions about how and why we are doing it. Or who we are doing it for.

Know your audience. They are the reason the workshop exists, and *their* objectives are what really matter.

Learning your ABCs

Educators have placed activities that are involved in learning into three broad categories. Today these are commonly referred to as affective, behavioral, and cognitive (or the ABCs).

The affective domain relates to how people feel about something, or, perhaps more precisely, the attitude and values they hold related to the content. Behavioral includes skills and performance they must practice or apply. Cognitive refers to knowledge – information and understanding that are required prior to mastering a skill or changing a behavior.

Early in my training career, I was fortunate to read Mel Silberman's classic text *Active Training*[iii]. His description of the ABCs of learning objectives has framed my approach to learning design ever since. Applying the ABCs to your web workshop's objectives means writing them in a way that expresses what your students can do as a result of their newly acquired attitudes, skills, or knowledge.

Note the emphasis on *doing*: instead of the student *telling* you they understand, the objective must illustrate the student in some type of post-instruction activity or behavior.

Any time we want our learners to develop a new skill, we should attend to affective needs, cognitive needs, and behavioral needs (usually in that order). As an example, suppose we suspect that our learners might be cautious or resistant to learning a particular skill. An affective objective would refer to the progress they've made from resistance to acceptance, or from caution to enthusiasm.

Similarly, a behavioral objective would target how well they've improved a skill or behavior (e.g. facilitating a virtual training session), and a cognitive objective would target their increase in knowledge (e.g. quickly identifying learning styles of virtual classroom participants).

If we want to teach someone how to deliver an interactive web workshop, we might use these sample objectives.

A = Affective	Rate the effectiveness of five virtual learning exercises that have been re-purposed from the traditional learning environment.
B = Behavioral	Deliver a 20-minute web workshop using all of the interaction tools.
C = Cognitive	Explain the uses and benefits of seven interaction tools.

Identifying objectives

An understanding of the ABCs of learning objectives is an important starting point. We now know that the purpose of our class objectives is to target the growth of attitude and feeling, behavior and cognition. But how exactly do we tell when development has taken place?

Traditionally we might give a test. It might identify whether participants learned the cognitive information. Of course, it won't work as well with attitude or skills. And as we'll discuss in Chapter 7, test performance in a classroom setting often weakly relates to how the student uses cognitive understanding *to perform a task* in the workplace.

Instead of asking students to repeat the learning points that you have made throughout the session, ask yourself the question, "How will I know whether participants learned the cognitive information or mastered the skills?" Be creative, and think in terms of what we touched on earlier: an emphasis on showing and doing, not telling.

Another good question related to cognition is, "What do learners need to *know* to perform well on the job?" This emphasizes the importance of knowing your audience, which we'll examine in the second part of this chapter.

How about attitude or emotion? If the topic of your workshop is related in some way to overcoming anxiety (e.g. public speaking), ask yourself if fear reduction is a valid objective and, if it is, how that would be expressed. Even for subject areas that wouldn't generally produce strong emotion, think about your learners' attitude. Ask, "How might learners feel about the content of the web workshop (Eager? Resistant? Scared?)"

Other questions that I've found to be effective are, "What tools and job aids would help learners be successful?" and "What skills do learners need to demonstrate to perform well on the job?" Or you might ask a question that incorporates the five basic elements: *who* will *do how much* of *what* by *when*?

Writing objectives using STAR

In the previous section we thought of some potential learning outcomes that represent a change in attitude, cognition or behavior. Now it's time to wordsmith them.

A good guideline is to write objectives that are specific, tangible, attainable, and results-oriented (use the STAR[iv] acronym as a memory jog). Here are some definitions and examples to illustrate this.

Specific. When an objective is specific, it is described clearly enough that there is no ambiguity about what success means. Include explicit conditions, like the number of times an action should be performed and/or how well it should be performed. Here are some examples:

> Poor: Begin to incorporate better practices in your web workshops.
>
> Better: Incorporate five interaction tools appropriately in the design of a web workshop.

Tangible. Tangible objectives contain concrete details that focus the participant on the end result.

> Poor: Understand how to build peer-to-peer collaboration.
>
> Better: Engage peers in collaborative problem-solving exercises during a web conference.

Attainable. Objectives must be within reach of the participant who completes the specific workshop. Stretch goals are fine, but if they're impossible to achieve, the objective becomes meaningless.

Poor: Get rave reviews on your next web workshop.

Better: Deliver your next web workshop with high levels of participation and interaction.

Results-oriented. Objectives must address an outcome, not a specific activity.

Poor: Practice polling, chat, whiteboard, and annotation to see how they work.

Better: Incorporate polling, chat, whiteboard, and annotation tools in the design of a 20-minute web workshop.

Finding the right words

Note the verbs that were used in the sample objectives. To help you pen STAR objectives, avoid tired, vague words like "learn" or "understand." If I'm coaching people on how to give a speech, how would I know if they reached this aim if they didn't actually give a speech?

Instead, use solid action verbs that are descriptive: explain, describe, apply, solve, create.

NetSpeed Note

If you're looking for strong verbs to accompany objectives, check out Designing Effective Instruction by Gary R. Morrison, Steven M. Ross, and Jerrold E. Kemp[v]. The book includes a "shopping list" of almost one hundred verbs that can be applied to objectives in the cognitive domain.

Our NetSpeed Learning's *Virtual Facilitator Trainer Certification* workshop culminates with students giving their own 20-minute training session. Here are samples of objectives they've written for their web conferences. As you read them, note the strong verbs. See if you can place the objectives into one of the ABC categories.

Topic: Increasing Achievement Drive (written by Tonya)
By the end of this web workshop, you will be able to:

- Identify the power of your own personal achievement drive.
- Describe best practices for increasing achievement drive.
- Write at least one effective motivational goal designed to obtain more of the rewards you want.

Topic: The Promising Conversation (written by Sue)
By the end of this web workshop, you will be able to:

- State the five possible responses to a request
- Explain the five elements of a reliable promise
- Name the five steps to securing commitment

Topic: Using evaluations to re-energize your board (written by Dave and Fred)
By the end of this web workshop, you will be able to:

- Articulate the benefits of conducting self-evaluations for board members
- List criteria that you would like to include in a self-evaluation
- Describe a procedure for utilizing self-evaluations and board evaluations to re-energize your board

Connecting objectives to exercises

It's best to focus on learning outcomes in this sequence: A then C then B. That is, influence your participants using an affective learning experience tied to an affective learning objective; give them content that will arm them to improve (cognitive); then provide the opportunity to practice it (behavioral).

Our challenge, then, is to design exercises and content to match the objectives. Depending on the ABC domain in which an objective resides, one instructional method may be more appropriate than another. For example, a lecture is fine for delivering cognitive knowledge, but less effective at changing attitude or improving motor skills. Here's one way you might relate various instruction methods to the ABC learning domains.

	Domain		
Method of Instruction	**A**	**B**	**C**
Lecture/presentation			√
Modeling		√	√
Role-playing		√	
Peer collaboration	√		√
Facilitator/Peer discussion	√		√
Video	√		√
Homework reading	√		√

Attending to Multiple Learning Modalities

Besides the instruction method, tailor your exercises to include multiple learning modalities: visual, auditory, or tactile learning, to increase long-term retention.

The belief that individuals have individual learning style preferences has been challenged by neuroscientists. The reality is that most people gather information through their visual sense (it's our dominant sense). John Medina, the neuroscientist who wrote *Brain Rules*[vi], states it best: "Vision trumps all other senses." Learners value pictures and diagrams to understand concepts. They'll respond best if you illustrate ideas with pictures or videos. They are attentive to the facial expressions and body language revealed by speakers in any video you show (as well as the expressions you reveal in your streaming image!).

But recognize that conveying information aloud adds another layer of connection through the auditory modality. Give participants time to actually read the text on a slide to themselves. Create opportunities to take part in discussions and debates, and ask people to hear and tell stories.

Add opportunities for tactile (or kinesthetic) learning. Learners often absorb concepts through hands-on practice and imitation, and benefit from demonstrations and labs.

Using all of the interaction tools to engage learners in exercises allows you to leverage these three learning modalities. I've taken a few informal polls over the years from my learners, asking how they benefit from different tools. Here's what they have told me.

Interactivity Tool	**Visual**	**Auditory**	**Tactile**
Chat	Medium	High	High
Polling	Medium	Very Low	High
Presenter's video	Very High	Medium	Low
Behavior model on video	High	Low	Medium
Whiteboard/annotation	High	Very Low	Very High
Status icons	Medium	Very Low	Medium
Audio	Very Low	Very High	Very Low

Wrapping up

To paraphrase one of Stephen Covey's[vii] habits of highly successful people, "begin your design with the end in mind." The sequence we've described so far does just that. Before you design an instructional sequence, consider the instructional method (e.g. lecture, poll, chat) and the learning modality (visual, auditory, tactile) that best meets the well-written objective, which you've shaped with ABC categories in mind.

And, as discussed, before you even begin creating objectives you should hear from your prospective participants.

Process	Action
Audience	1. Begin with an understanding of your participants' real-world challenges, skill gaps, experience and concerns.
Objectives	2. Determine learning outcomes. 3. Write objectives using STAR and strong verbs. 4. Review your objectives to see if they reflect the ABCs.
Presentation	5. Select the instructional method (e.g. facilitated discussion). 6. Develop PowerPoint slides.
Interaction and Collaboration	7. Create specific exercises using the interaction tools.
Pilot	8. Test the design with a pilot audience.

Knowing Your Audience

We touched on the importance of writing workshop objectives that answer real-world challenges, concerns, and attitudes of the learners who will attend your training. Now that the objectives are defined and the workshop has been designed, the remainder of this chapter will focus on your attendees.

Remember our virtual classroom example in the previous chapter? It illustrated the importance of creating a warm environment for your learners. The variety of interactive techniques used in the example helped Gerri feel comfortable and involved.

The more you know your audience – their names and faces, their backgrounds and skill sets, the culture and politics of their workplace, their reasons for showing up to the session – the easier it is to not only engage them, but to prepare and deliver your presentation in a way that reaches your workshop's objectives.

You'll get to know your learners more and more as you interact with them. But the time to solicit information about *their* learning environment and expectations is before the workshop begins. Let's explore that further.

Getting personal before the workshop

Prior to the start of your workshop series, it is common practice for participants to receive an email or calendar notice containing course objectives, schedules, and perhaps a biography of the instructor. This is an opportunity for getting to know your learners. Why not follow up the email or notice with a phone call so you can meet a few participants on a more personal basis? Or offer a pre-assignment that invites them to post responses to a discussion thread or class blog.

If possible, ask detailed questions that can help in your preparation for the class. For example:

- Their place in the organization: their job title, roles, skill sets.
- Challenges and opportunities that the learning sessions might address, with specific examples.
- Learning goals for the class.
- Demographics: male/female, generational differences, seniority or experience, education levels, comfort/familiarity with technology, other demographic information that pertains to the topic of the web workshop.

This information may prompt you to modify your workshop objectives to meet their learning needs. It may also impact the amount, type, and duration of interaction and collaboration exercises.

Before each class of your workshop, contact your learners again to set expectations. Send handout pages and give an easy assignment to be completed before class. This will fix their focus, and sets the stage for their involvement during a class exercise.

Do not make the mistake of sending out the slide deck, even if requested. Participants often request it as a note-taking aid, but there is no advantage to this; the disadvantage is that if they think the slide deck captures the important elements, they may be tempted to only listen with half an ear.

Getting to know your organization

I stated in the opening of this book that the shrinking size of our 21st century world is heightening interest in virtual training as an alternative to the physical classroom. With change comes challenge, ones that involve technical aspects, as well as policies, politics, and culture.

One of our prospective customers, a discount retail giant, wanted to deliver web training to all of their store employees. But their in-store computer network was used for inventory and cash registers, and they didn't have the capacity to stream a web workshop without seriously compromising their ability to serve their customers. Imagine the wasted time and tarnished relationship that would have resulted if we hadn't discussed this ahead of time.

Another NetSpeed Learning client used VoIP voice technology for its workshops. With VoIP, audio is heard over a computer speaker instead of a phone. Though each participant was equipped with a VoIP microphone, they were reluctant to broadcast their thoughts because the audio blared loudly from their speakers, which tended to disturb their co-workers.

After digging a little deeper with management, it became clear that a top-down policy decision had been made, and a lot of money on technical equipment had been spent. Instead of an Internet environment that promoted interactive learning and collaboration, the organization had unwittingly embraced a solution that delivered the opposite. The moral: know what you're getting into and help influence it if possible. At the least, I came away from the conversation knowing how to tweak my web workshop to anticipate limited audio responses.

Don't launch your web training without knowing how the organization has trained its employees traditionally. If physical classrooms have been the norm, be ready to counter the negative expectations of Chapter 1. If you already work for the organization, you are the best advocate your company can have.

If your company has embraced eLearning (which would include any learning delivered through the Internet or the company's network), expect less of a technical ramp-up and be prepared to emphasize interaction and communication. If interactive web experiences are old hat, you'll know to vary your warm-up exercises and the session pace accordingly.

A few best practices

We've mentioned that getting to know your learners ahead of time will return huge benefits when it comes to strategies on how best to create a warm environment and sense of community. Other advantages deal with more mundane, yet important, logistical decisions.

Here are some tips and best practices.

- Review the participant list and learn their names prior to the session. Use people's names throughout your web workshop.
- Ask for photos of your participants ahead of time, to get a sense of the individual. Have them upload their photos to a social learning platform or website. You'll also be able to utilize these photos to help your participants get to know each other better. We'll talk more about that in Chapter 4.
- In Chapter 2, I recommended a small class size for the virtual classroom. As far as the length of the class goes, try to keep it from 60 to 90 minutes. It's hard to sit longer than that in front of a computer screen. If there is little interaction between you and the participants, as in a marketing webinar, keep the class length to a maximum of 60 minutes.
- The class length should also be tailored to your participants' roles and job titles. Don't expect the sales team to sit still for long periods of time. Customer Support personnel also have limited time. Ask questions up front to get a feel for this. In addition to class length, you may get an inkling of how scattered or focused your learners are likely to be.
- The best time of day for a web session mirrors the traditional classroom setting. The morning hours through lunchtime are often best; avoid the afternoon right after lunch when your learners are at their lowest level of energy.

- That said, remember that you may be presenting to people in different time zones and in different countries. NetSpeed Learning Solutions is in Seattle, so mornings in Washington State work well for North American attendees. With the increase of global training, you might have to compromise on schedule to reach your audience.

Talking about time zones tends to reinforce one of the wonderful aspects of being a virtual facilitator. Your LinkedIn contacts will expand with names and faces from all over the world.

I'm reminded of a virtual session I delivered for the ASTD International Conference.[viii] It was scheduled for 7:30 a.m. Eastern time. This is an ungodly hour for East Coasters, and unconscionable for those of us from the other coast, especially those of us who are seriously jet-lagged from arriving the night before.

During the intro, I used the topic of time and the sleepy state of the participants (and facilitator) to get to know everyone better. It turns out that the time of day was just right for one of my audience members. This was the fellow from Finland, who had logged in at 2:30 p.m. in his time zone!

Homework

Assignment 3-1: Choose a web workshop topic and write objectives

The expectation is that by the end of this book, you will have designed (and perhaps given) an effective virtual classroom exercise using web conference interaction tools. Choose a topic and prepare 3 learning objectives for your 20-minute web workshop. Ensure that these objectives accurately describe the outcomes/results you expect for learners at the end of the workshop (what will they be able to do?). For example "Coach others to correct performance problems" is focused on an outcome. "Understand the steps needed to coach someone" is less specific and less focused on the outcome.

[iii] Silberman, Mel with Carol Auerbach. Active Training: A Handbook of Techniques, Designs, Case Examples, and Tips. San Francisco: A Wiley Company, 1998.

[iv] The STAR acronym was created by NetSpeed Learning Solutions for *Managing Projects by Design*, a module in the NetSpeed Leadership blended learning program.

[v] Morrison, Gary R., Steven M. Ross, and Jerrold E. Kemp. Designing Effective Instruction. John Wiley and Sons, 2007.

[vi] Medina, John J. Brain Rules. Pear Press, 2014.

[vii] Covey, Stephen. The Seven Habits Of Highly Effective People. New York: Simon and Shuster, Inc., 1989. One of Covey's habits is, "begin with the end in mind."

[viii] The ASTD International Conference is given annual by American Society for Training and Development. http://www.astd.org/

chapter four

interaction and collaboration

We've already mentioned that the most boring web workshops feature the presenter droning on about the topic without engaging participants. Without opportunities to interact with the facilitator and each other, students rapidly discover how easy it is to multi-task or even leave their desks to run a few errands down the hall.

Web conference platforms allow for interaction using a variety of tools, like chat, polling, whiteboards, application sharing, annotation tools, break-out rooms, and document sharing. The best facilitators are able to use the features of their web conference software creatively to engage their learners, encourage collaborative problem solving, share ideas and opinions, and increase peer-to-peer learning.

In this chapter, we'll take a deeper look at how to promote interaction and collaboration, using the tools built into most web conference platforms.

Defining interaction and collaboration

First, let's define our terms. In a virtual classroom setting, "interaction" refers to verbal and written synchronous (two-way) communication, usually between the facilitator and the participants. The term "collaboration" refers to the act of working together to achieve a goal, and often has the participants contributing independently from the facilitator.

When interactive exchanges take place, there is a clearly defined leader, the facilitator, and the learning is fairly linear. The facilitator makes use of questions or polls to check understanding, maintain interest, promote the sharing of ideas and experiences, or otherwise engage participants in learning activities. All these activities, of course, drive the session toward meeting the learning objectives.

In a collaborative exercise, a communication exchange still takes place, but there is more emphasis on having learners actively work with each other toward an end goal. The goal might be as varied as creating a document, making a decision, completing an exercise, or giving a presentation. In this way, they leverage each other's knowledge, not just what is gleaned from the facilitator.

Besides web platform tools, another way to collaborate is through the effective use of social media tools such as class blogs. The previous chapter's discussion on learning objectives stressed that we should be less interested in having learners pass tests (or get their homework graded) than in actually demonstrating learning by performing the learned function or skill. In keeping with that philosophy, it's a wasted opportunity for students to submit their homework to the instructor to show that they've completed an assignment. Instead, a savvy instructor might have students post their homework on a social media site available for view by all the class participants. Not only is the wisdom shared, but by encouraging learners to comment on each other's work, interweaving threads of feedback allow for a sort of asynchronous collaboration.

In chapter 7, we'll discuss social media tools in more detail.

NetSpeed Note

The full definition of synchronous learning, as we'll be using it in this book, is "Any scheduled learning event where interaction occurs in real-time." With synchronous learning, interaction between the facilitator and the participants is not delayed over time, as it would be with asynchronous learning.

Examples of synchronous learning are the traditional physical classroom, a mentor and client communicating over the phone, and a facilitated virtual training session. Examples of asynchronous learning include email exchanges, microlearning, a course hosted on a social learning website, and self-paced eLearning.

You'll find definitions of these terms and more in the Glossary section at the end of the book.

Why the distinction matters

Interaction and collaboration are similar in many ways. But the distinction between them matters, because you'll structure your learning segments and exercises around them.

In my opinion, your training session will benefit from interaction between facilitator and learners in almost any situation, since its primary purpose is to foster engaged participants. Compared with interaction, collaboration exercises often consume more time and because participants are driving the dialogue, the facilitator has to be willing to release control over the outcome. In addition, some of the tools take longer to master. This doesn't mean you shouldn't use collaboration, it just means you'll want to design your sessions intelligently, using both interaction and collaboration where they make the most sense.

Here are some questions you can ask yourself to help design your session around interaction and collaboration.

What is the purpose of the training segment?

Is the program or exercise's intent to develop the art of decision-making, alternate approaches to a challenge, or coming up with solutions? These topics all benefit from discussion and sharing of expertise. Incorporate collaboration if the intention of the exercise is to engage learners in peer-to-peer learning.

Are the learners liable to have domain knowledge that would help their fellow learners?

Of course, if your participants don't have the right skills and experience, sharing and interaction may have the benefit of engaging them (the primary purpose of interaction tools) but not be as useful in leveraging the individual's subject matter expertise (the purpose behind collaboration). The more advanced the topic, generally the more likely that collaboration will be effective.

How do your learners learn?

It's helpful to assess how your learners learn. Interactive sessions can be more structured, with direct presentation of materials. A collaborative session requires much more participation, especially of participants who have expertise that they are willing to share.

How can I design for collaboration if I don't know these things about learners ahead of time?

One approach is to gather information about your participants – their communication style, their subject matter expertise, their willingness to share – by way of interaction (polling, hand raising, and discussion). Then use that information to decide whether to use collaboration, and how to design the exercises to greatest advantage.

If your workshop is composed of a single session, you may want to prepare two alternate ways to go in advance. If it's a multi-session workshop, you can apply collaboration to future sessions. In that scenario, you also have the luxury of determining your learners' characteristics through homework exercises.

Using interaction tools during participant introduction

Let's show the interaction tools in action. Our participant, Gerri Jordan, is now engaged and actively participating in the *Virtual Facilitator Trainer Certification* sessions. For today's class, she's already downloaded and glanced at the eight-page handout that they'll be using. She knows that while Communication Styles is the nominal topic, the real intent of the session is for the instructor to model a sample of web conferencing tools to promote interaction and collaboration.

Gerri sits in front of her PC with her headphones on, a few minutes early for the session. She's studying the handout when Cynthia's voice is heard.

"Hi, everyone. Welcome and thanks for attending!"

"Hi, Cynthia," Gerri says. Her greeting is echoed by the other students.

As Cynthia's image is streamed and the first slide is displayed, she runs through the familiar housekeeping issues, then introduces the day's goals.

"Today's session is sort of a hybrid. You'll be participating as though you've signed up for a *Working With Communication Styles* class. Of course, the main goal is to learn interactivity and collaboration tools that are available from our web platform software. From time to time, I'll step out of Communication Styles facilitator role to comment on the tools, and when and how to use them. Got it?"

Sounds of understanding and agreement.

"Good. You used chat, polling, and raise hand in our previous sessions, and will use them again today. And we'll introduce

> you to a few more. I may also be calling on a few of you by name. But now, I'll put on my Communication Styles facilitator hat."

This should be fun. A little schizophrenic, but fun.

> "You're probably all familiar with the concept of Communication Styles. Today we'll increase your awareness of these styles, and increase your acceptance and appreciation of style differences and how to adapt them to achieve good communication results."

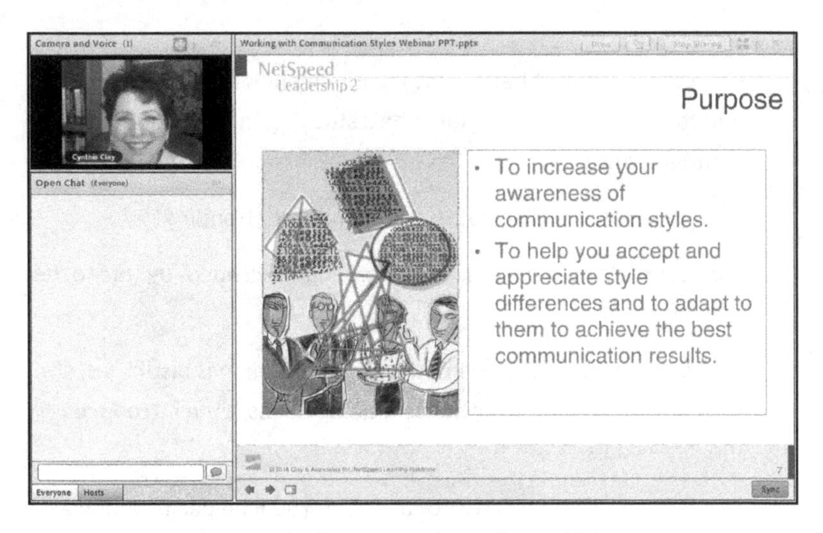

Cynthia goes over the housekeeping rules, which are by now familiar. Then she discusses the learning goals. After that, she switches to a new layout and displays a chat box.

> "I've opened a new layout. Feel free to offer an example of a communication style behavior that's different from your own. Just think about some way of expression that is *not* naturally yours. I put a couple examples at the bottom."

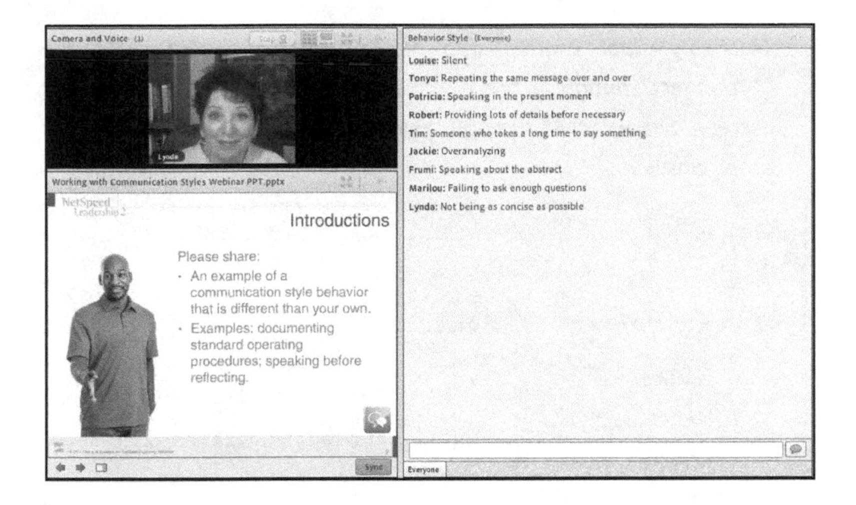

As the chat window starts filling up, Cynthia verbalizes what she sees, with some short comments.

"Louise, you say that silence is not your natural preference. Tonya, a behavior that is different than your own is repeating the same message. I also see behaviors such as speaking in the present moment; providing lots of detail before necessary. I'm like that myself…."

I jump to conclusions quickly, thinks Gerri. So I'll write something down about analyzing.

"Gerri, you're observing that the behavior 'to overanalyze' is not yours. Interesting…" Cynthia reads all the chat messages, then says, "So we're definitely aware of characteristics about others that we don't share. Next we'll do some opening exercises. These will introduce some basic differences between people."

A new layout appears with a slide that asks for Birth Order.

"I'll move the poll pod to the center. What order were you guys born in?"

Just my older brother and me, Gerri thinks, and clicks the "Youngest" option in the poll window that has appeared on her screen. She watches the poll results grow as her classmates enter their answers.

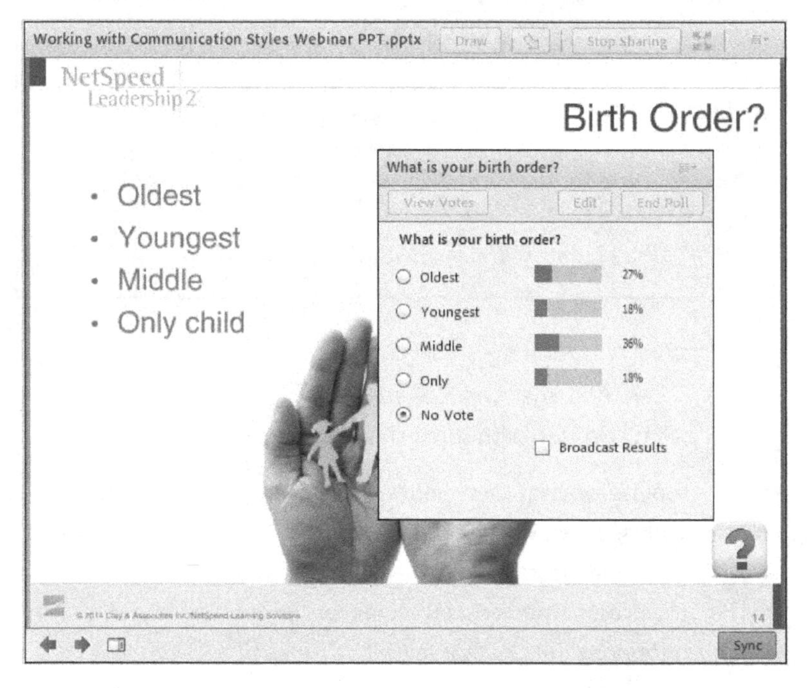

"Look at this, 36% were middle children," says Cynthia. "Just two only children, and two youngest kids. I'm curious: who are the youngest?"

That's me, thinks Gerri and unmutes her phone. "I am. This is Gerri."

"Me, too," says another voice. "This is Patricia."

"You're the youngest!" says Cynthia. "I'm surprised because you've got such a take-charge quality about you."

"I'm surprised too," says Gerri. Oops, she's said it out loud and hasn't muted her phone.

Cynthia laughs, and so does Patricia. Gerri mutes her phone, still chuckling.

This is really incredible, she thinks. Not only are we participating, but I actually feel like I know some of the others.

Cynthia continues displaying polling slides: where you were born, what color eyes and hair you have. With each question, she laughs, comments, asks certain responders for more information. By now most people are unmuted, and the conversation is rollicking.

Cynthia asks everyone to mute their phones again, then sums up what they've learned by saying that these are traits that are hard-wired, difficult or impossible to change, part of our nature.

> "But what about our communication styles?" Cynthia asks. "You've all read the handout and perhaps have started to figure out your own communication style. Do you think your communication style is hard-wired too? Let's see a show of hands."

Using chat, poll and raise hand for participant introductions

In the example above, you can see how I've used interaction tools to promote engagement. Remember, that's one of their main purposes. At the beginning of the session, this interactive discussion serves the same purpose as the around-the-room participant introductions that take place in the traditional classroom. In a physical setting, introductions (or "mingles") are where we assign a name to a face. In the virtual setting, we need to rely on voice and information to assess our classmates.

The tools of chat, polling, and/or status icons (also called emoticons) like the raise hand tool are especially helpful for this important initial ice-breaker. Let's dissect the steps of the example to learn how we pulled off our mingle.

When you present an opening question in chat, the responses (with associated names) start to pop up. As examples of communication style behaviors appeared in the chat box, I read off both the participant name and contribution. "Jim says a communication style behavior that is different than his own is to be aggressive. Sue observes that others may be good listeners who prefer to ask a lot questions before giving their opinions." People love to hear their name spoken. And the others in the virtual room will slowly start forming a mental image of that person.

With polling, it's a good idea during introductions to ask a question with just a few response options. Make it something that can be easily answered and that tells something about the person without, of course, being too personal or intrusive. When possible, ask a question that can segue into the subject matter of the workshop.

The initial poll results are in percentages or raw numbers (or both), and that allows you to follow up with voice or chat to place names next to the responses. There's an added advantage to this. Because polls chunk up the respondents into categories, participants immediately relate to the other people in their group. In the example, Gerri learned something about a fellow learner that she'll most likely remember; they were both the youngest child in the family.

Besides chat and polling, we often use a status tool like raise hand in mingling. In the example, I used it to get agreement or disagreement on whether communication styles are hard-wired. Of course, besides warming up the audience, I'm segueing into an important distinction between traits that are part of our nature and communication styles. (Okay, I won't leave you hanging. The former are not easily changed; the latter is a combination of natural gifts and adaptive behavior, so they can be tweaked.)

Chat, poll and status icons are the "big three" because they're so handy in so many situations, not just for participant introductions. We discussed the raise hand tool in depth in Chapter 2. Let's do the same for chat and polling.

Letting it flow with chat

Here's one of the most valuable suggestions about chat: let it flow. Allow participants to chat freely about a topic with each other. Encourage them to comment on others' entries. If you ask an engaging question, you're more likely to encourage debate and bring up differences of opinion.

Sometimes you'll want to probe for deeper responses. Select specific people to unmute their phones to explain their responses.

As the example demonstrated, it's a good idea to speak aloud key points as entries appear, and use the participants' names. If you're quick, you can even refer back to their earlier points.

In small group chats, open multiple chat pods (or chat windows) for separate groups. Have them chat about the same or different topics. When the chat session is done, request a spokesperson from each group to summarize key points out loud (if using tele-conferencing or two-way VoIP), or summarize and connect key points yourself (if using one-way VoIP).

Be sure to "recognize and reward." Thank and praise the first learners to chat. Offer simple prizes, if appropriate. Notice when someone begins to chat later in the session and thank them for adding their thoughts.

Polling for knowledge or interest

Use polls to check knowledge or experience, stimulate interest, or set up a discussion. Polls provide instant feedback (and satisfaction), allow learners to compare their responses, and help the facilitator lead the discussion and tailor the lecture.

Engage the brain. Don't just present research results. Ask participants to guess or to estimate with a multiple choice poll. Then present the data and compare it to their poll results. Memory and retention are often improved by guessing incorrectly and then learning the correct answer. And, if someone isn't guessing, but is sharing what they know, their correct answer will then be validated.

If you're trying to quickly gauge the level of experience of your participants, use polls with Yes/No responses. True/False polls are useful for quizzing the class to create interest in the topic. You can also use them to check for comprehension.

If it is a poll about experience levels or personal facts, allow participants to see the poll results as they are entered in real time. If there is a right answer, share the poll results for each question after everyone has responded, so as not to bias their answers.

Having fun with collaborative learning

I touched on using chat in small groups. Some web platforms (Adobe Connect, for example) allow you to have multiple chat windows open simultaneously. This encourages debate and discussion among the participants. That's what we're after. Roger Schank, in his book *Lessons in Learning, e-Learning, and Training*[ix], says this:

> *"The elements of fun are the same for everyone... The underlying concept is engagement. In other words, if you are really into it, that's fun."*

Remember that collaboration is interaction with a purpose. Communication exchange still takes place in collaboration, often using interaction tools, but there is more sharing of expertise among the participants, often toward an end goal. Let's continue with our example to see how to use chat and other web platform tools to promote collaboration.

A few minutes have passed since we peeked in on the session. Cynthia has walked through a self-assessment exercise with the participants that consisted of ten word pairs. Each person selected the word from each word pair that better described them. For example, do they focus on Facts or Ideas? On the Present or the Future? Are they more likely to Imagine or Execute?

"Okay, now that you've selected the ten words that most describe you, circle them on the Communication Styles Model in your handout. It's the one that looks like this."

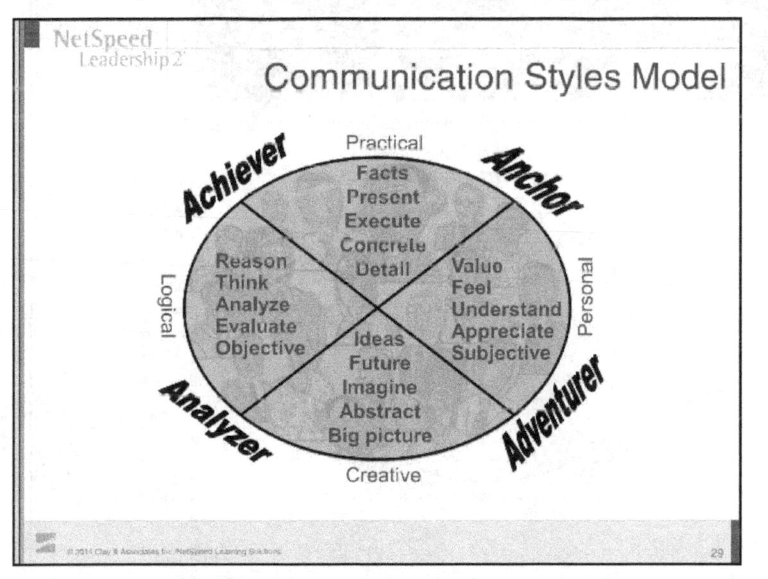

Wow, I've got practically all the words in the Personal and Practical sections circled. Which makes me an Anchor. Not sure what that means. But I have a feeling I'm about to find out.

And she's right. Cynthia runs through a series of slides that explains Communication Styles – what they are, how different communicators deal with different situations, their areas of strengths. As she has done throughout the workshop series, she changes slides every minute or so, and maintains interest by involving the group.

> "Raise your hand if you're comfortable in saying, 'This is my style'. If you're not comfortable, or you need to hear more before you decide, click the X. By the way, here's what my model looks like. I'm an Analyzer."

Gerri raises her hand. She sees that only Marianne is still not ready to go forward. Cynthia asks her to unmute her phone. It turns out Marianne just has a clarifying question.

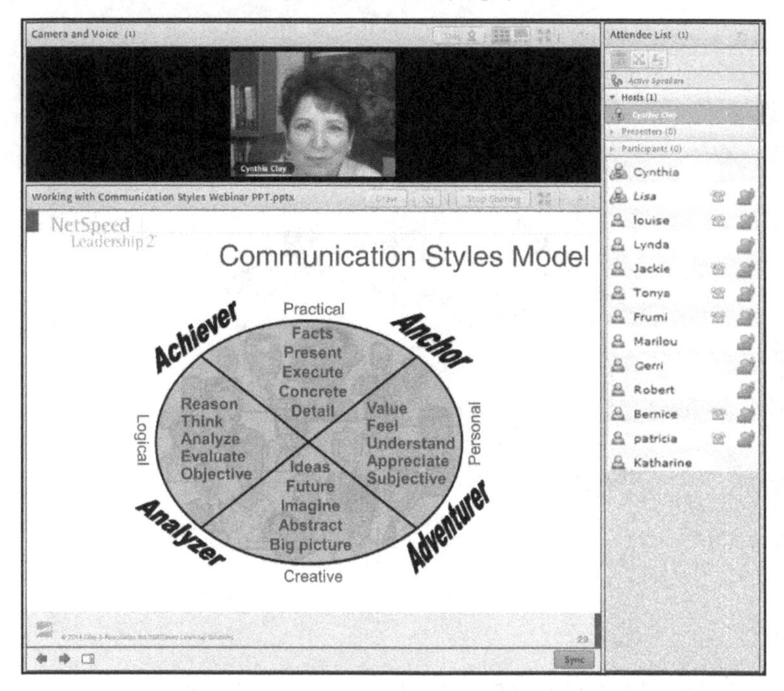

> "So let's see what communication styles you have. I'm going to take us to a new layout with a poll, and I'd like everyone to tell us your style."

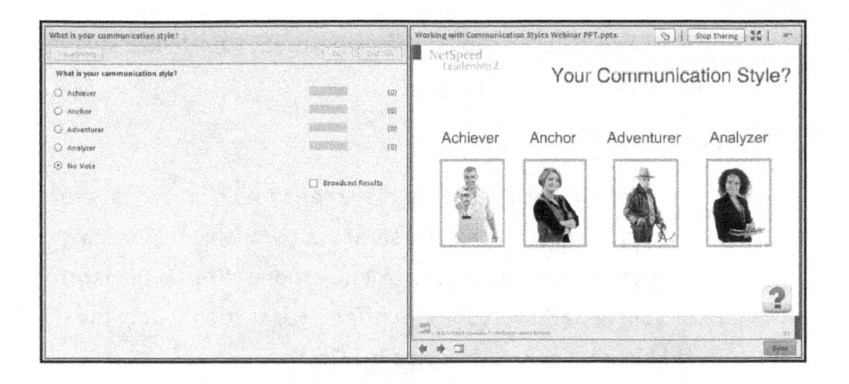

"Let's see. It looks pretty even so far. Now one style is breaking out. Okay, looks like we're all done, except one person."

"Probably Marianne."

"Okay, done. Here are your results. Anchors lead, the others all have two each."

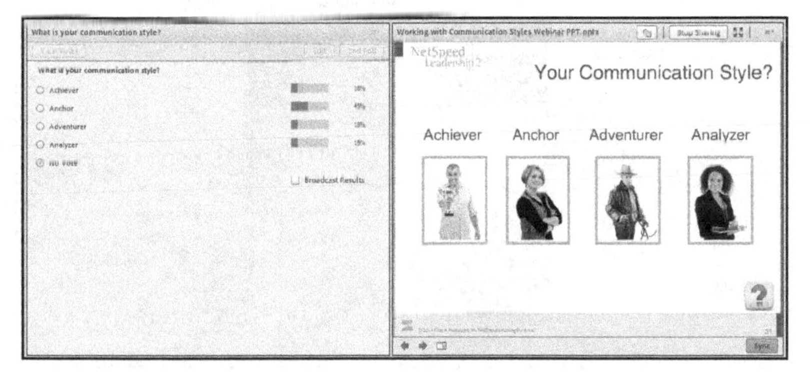

"I often have more Adventurers than anything else. Adventurers like the creative side of work; they get things done with a sense of adventure and innovation. But it also makes sense that Anchors are drawn to training. Good execution and attention to detail are important."

Looks like I'm in good company. I wonder who the others are.

> "Now I want to give you a chance to meet your fellow style-mates."

She read my mind.

> "We'll do this by dividing up the screen into chat boxes. Look at the questions on page three of your worksheet. For now, I want you to just discuss the last question, 'What is the motto for your style?' Everyone can offer a suggestion, but in the end just pick one. Got it? Ready? Go."

This is neat. Gerri types: "Hi, fellow Anchors." They discuss mottos. Tonya suggests 'Feeling trumps logic.' John offers 'Carpe diem'. But Gerri, in keeping with both the communication type and the theme of the web session, suggests 'Collaborate for common good.' The group likes it and picks it for their motto.

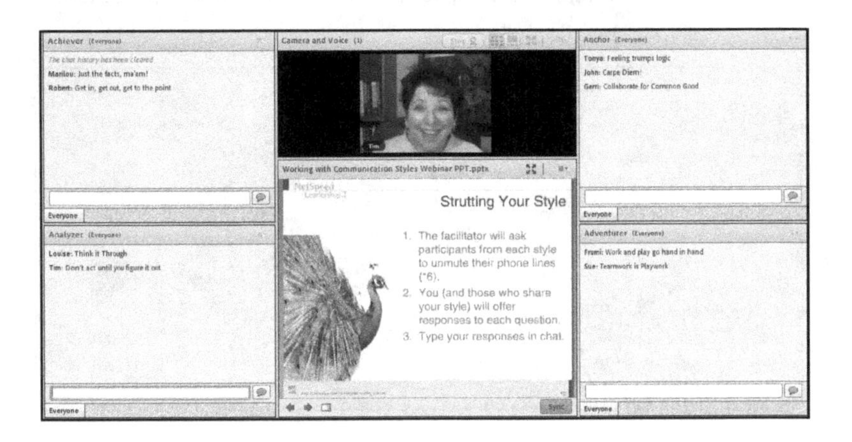

The chat windows fill up.

> "Okay, it looks like everyone is done. Thanks for such great discussions. I'd like one person from each style group to unmute your phone line and share your responses.
>
> "Another way we could have done this exercise is through break-out rooms. In a break-out room each group goes into a

private session where they can use any of the web platform tools among themselves, as well as speak to each other. It's really the best way for small groups to work together for consensus, but be aware that it takes some practice and if not done right, you can lose people in cyberspace."

That sounds serious.

"Now I'm going to give you a chance to strut your style."

A slide with an exercise is displayed.

"Go back to your worksheet. This time, complete all the questions individually. One by one, each group will unmute your phone and explain your responses. One person from each team – why don't we make it the person who suggested the winning motto – will type them onto the whiteboard using the annotation tool. This is yet another tool that you can use in collaboration."

A new tool! I haven't heard of this one before. Lucky me, I get to use it.

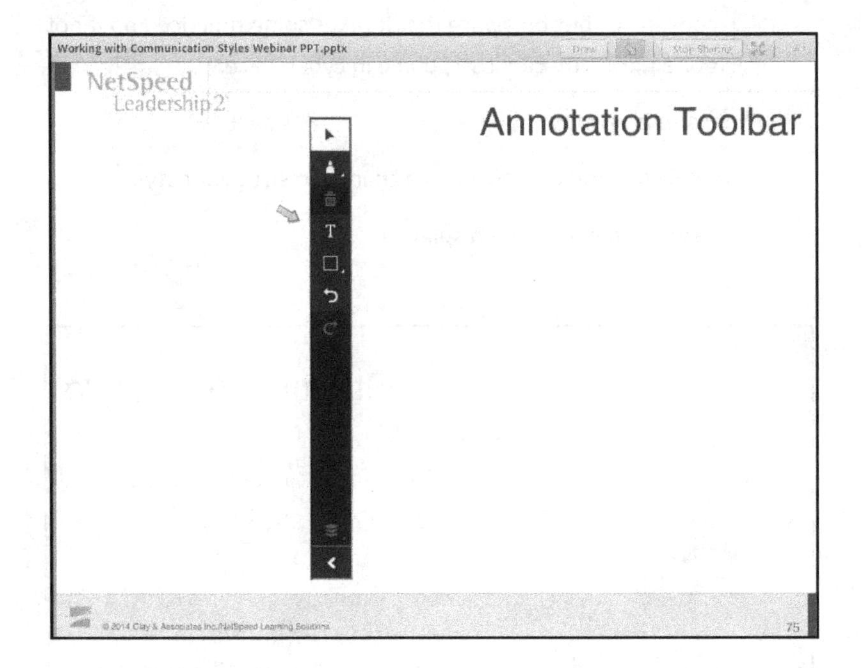

"The annotation toolbar appears to the left of your slides. When you click on the **T** tool, you'll see a font size."

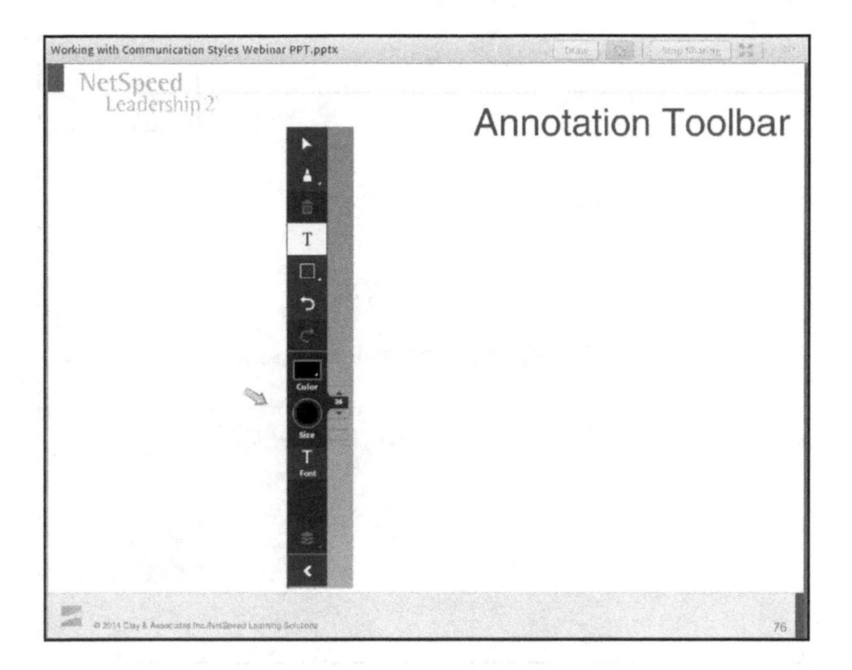

"Change the size of the font to 20 pt. We've found that this is the best size. Let's start with the Anchors. Ready, guys? Please unmute your phones. I'll put the questions on the next slide. Anchors, who's your scribe?"

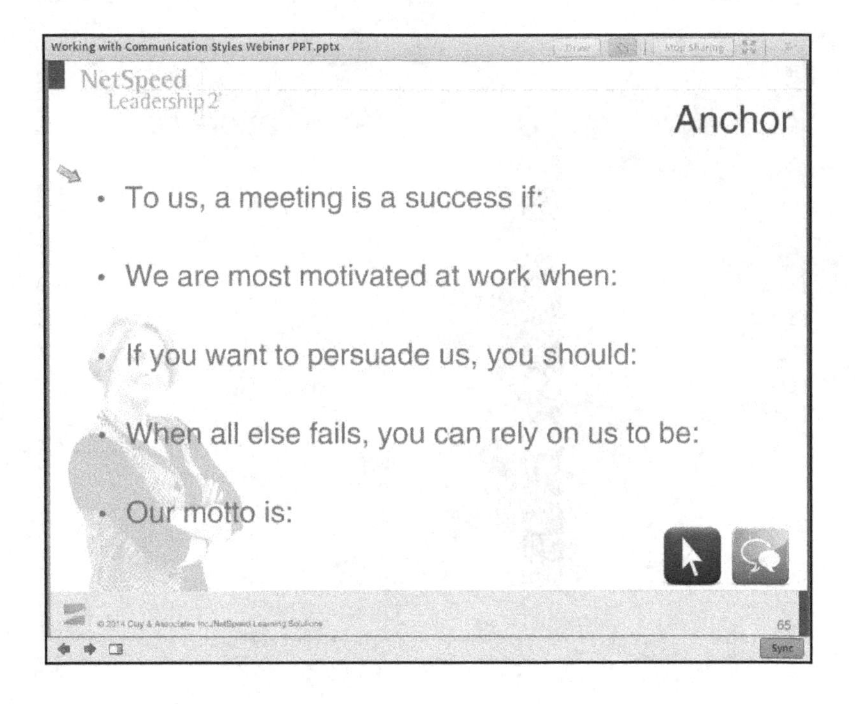

"It's me, Gerri."

> "Okay, good. Anchors, how would you answer the first question? How do you know a meeting is a success?"

"Everyone gets along," said John. "Well structured," offered Suzy.

"Share feelings," says Gerri. She continues to record all the answers on the slide, and they work their way through the questions.

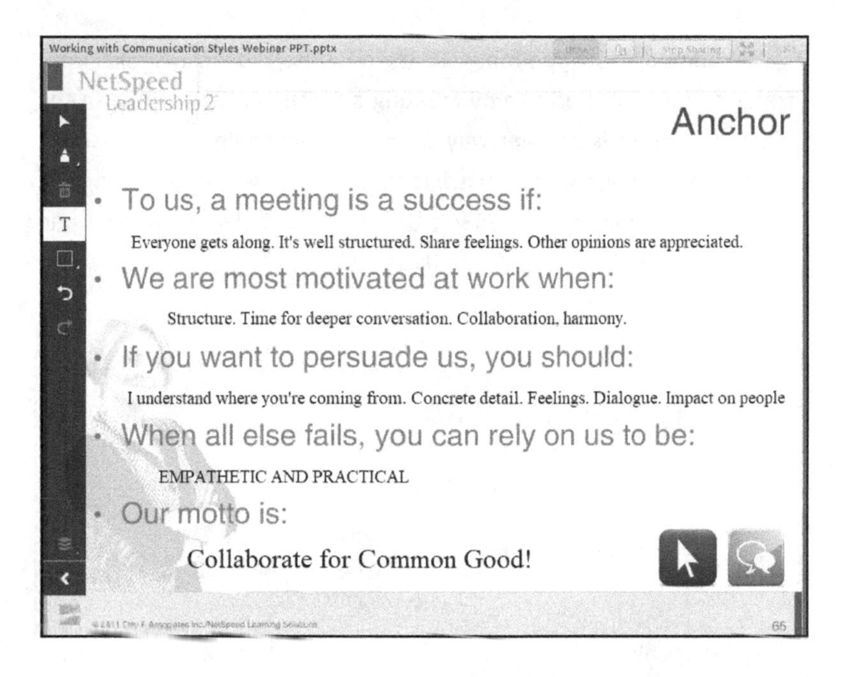

Finally all the questions have been answered but the last.

"And your motto? Gerri?"

"Collaborate for common good!"

NetSpeed Note
Look at the beginning of Appendix J for a checklist of questions to help you prepare to give a workshop similar to the one that Gerri just experienced.

Performing whiteboard wizardry

Gerri is getting good at using the annotation tool on the whiteboard. She used the whiteboard in a standard way – to record ideas in real time. Besides typing in the text, she could have also used free-hand drawings tools like lines, circles, and squares.

In addition to appointing a single scribe, you can make your exercise more collaborative by creating a "group grid." The one in the following graphic is another way to perform a mingle. Each square of the matrix is assigned to a participant. Ask learners a question (e.g. "Tell us one thing about yourself that not many people know") and have them type their response in their assigned square.

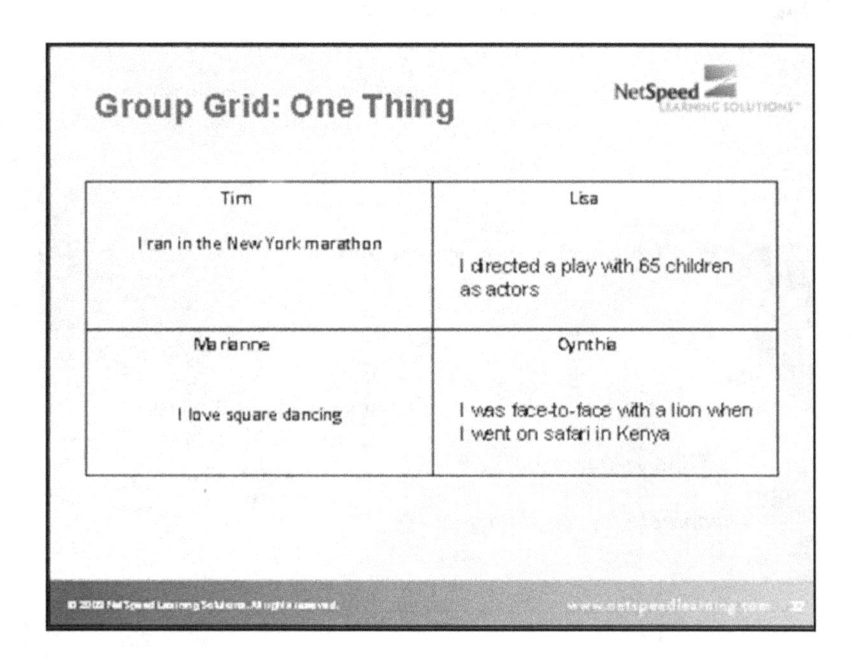

A variation of this exercise uses pointers. Instead of assigning your learners a place in the grid, have them claim their spot using their pointer.

Then there's the "Team Challenge." Ask for five volunteers to complete a challenge to draw an object or animal together on a whiteboard. Explain that they can organize their process first but once they begin, no one may speak. Watch the picture develop – it's usually as hilarious as the sample shown below from an actual session. Debrief by asking probing questions about teamwork, so the team reflects on what they've learned by doing and the observers share their insights about teamwork.

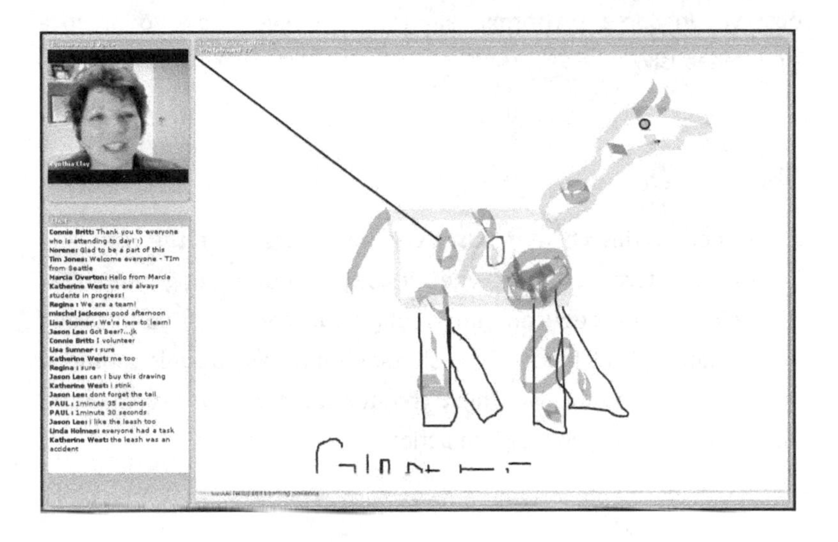

Sharing desktop applications

Every web conferencing tool includes the ability to share applications that are on the presenter's desktop. In some applications, that's how you'd display PowerPoint – by pulling up a slide deck that exists on your desktop.

(By contrast, Adobe Connect, WebEx, and other web conference platforms allow the presenter to display a PowerPoint within the platform itself. You can upload the file from your desktop. Usually this way of displaying PowerPoint allows for easier integration with the interaction tools.)

Besides PowerPoint you can share other applications – word processing, spreadsheets, graphics – as needed. You can also access other websites to demonstrate, for example, how to use a software application.

Be aware that you can also turn over control of the session to specific individuals among your participants. This is very handy for giving them the opportunity to practice steps or actions while others observe. In some platforms, all users are even able to practice simultaneously.

Showing tool usage and benefits

Hopefully this chapter has given you a taste for the uses and benefits of interaction tools when it comes to promoting interaction and collaboration between and among your learners. If you use the tools creatively and liberally, you shouldn't have any trouble capturing and keeping attention. The next chapter will give you several more opportunities to see the tools in action.

The following table summarizes the uses and benefits of the tools. It's repeated in Appendix B.

Tool	Uses	Benefits
Chat	• Solicit learner input • Encourage collaboration	• Actively engages learners in discussion • Creates peer exchanges
Polling	• Check knowledge or experience • Stimulate interest • Set up lecture or discussion	• Provides instant feedback (and satisfaction) • Learners can compare their responses • Helps facilitator lead discussion and tailor lecture

Tool	Uses	Benefits
Status Icons	• Quickly get input • Identify volunteers for exercises • See agreement or disagreement	• Participants can "vote" or respond though they may be uncomfortable using chat • Opens the door for facilitator to call on learners to give examples
Streaming Video	• Streams video of Facilitator • Adds animation and interest	• Helps to establish rapport • Creates a sense of connection
Whiteboard/ Annotation	• Brainstorm and capture ideas • Encourage collaboration	• Allows facilitator to guide and record discussion visually • Encourages peers to share ideas
Application Sharing	• Share websites, your desktop, or documents • Can turn over control to specific participants	• Allows you to demonstrate steps or actions online • Gives a participant the opportunity to practice steps or actions while others observe
Breakout Rooms	• Have participants work in small groups	• Supports practice and feedback • Encourages quieter participants to participate verbally

Homework

Assignment 4-1:

Interactive Exercises for Your 20-minute Web Workshop

Develop six interactive exercises for your web workshop. Use at least one of each of these tools: polling, chat, status icons, and whiteboard. Your document should describe each exercise with enough detail that another facilitator could present it.

[ix] Schank, Roger. Lessons in Learning, e-Learning, and Training. San Francisco: John Wiley and Sons, Inc., 2005, p. 209.

chapter five

powerful PowerPoint pointers

We know that there are more examples of bad PowerPoint slides than good ones. Why is this rich, powerful program used so poorly? And do best practices differ between traditional and virtual learning sessions?

People learn and process information differently, using different senses. For better learning and memory, we want to mix modalities, so people hear the spoken word, read text, see graphics and animation, and apply content hands-on. Good instructors appeal to more than one sense. They also appeal to their listeners' emotions, since if you transfer knowledge using emotion, it will be branded more deeply into memory.

Seth Godin in his e-book *Really Bad PowerPoint*[x] emphasizes this last point. He encourages us to "make slides that reinforce your words, not repeat them. Create slides that demonstrate, with emotional proof, that what you're saying is true, not just accurate."

In other words, don't just recite text that your participants can easily read themselves. Godin recommends that you use an emotion-inducing graphic, or (for example) the actual series of frustration-inducing steps of your company's inadequate software application. While your viewers are looking at images on the screen that represent a problem, your words are selling them the solution. (There are caveats to this advice that we'll discuss further along in this chapter.)

It's a worthwhile exercise to attend web presentations and I always ask my *VFTC* students to spend a few minutes developing their personal list of top ten PowerPoint strategies to share with the class. They've come up with some great suggestions.

Improving slide design

Many of their comments – a full 40% – turned out to be basic design points. That's to be expected: the overall aesthetics of a slide sends a message to our brain before we even begin to try to understand the content. Comments like the following have been offered.

Use colors that have a good contrast, but remember that plain B&W is boring. ~ C.J.

Use bold visual designs. Follow the rule of thirds in which graphics do not appear predictably in the middle of the slide. ~ Lynn

Be very selective about the clip art used. We've seen most of the Microsoft graphics. ~ Dave

Use animation if it's needed to make the point. ~ Vernon

Besides aesthetics, single-slide designs need to ensure that the viewer can understand the individual elements that make up the slide.

Fancy, flowery, Hallmark type fonts are hard to read and sometimes annoying. Best to avoid them. ~ Carole

If you are using graphics, make them large enough to make a point, otherwise why bother? ~ C.J.

Font size and color is important to the audience. Nothing is more distracting than a slide one can't read because the type is too small or in a color that won't transmit in the room. ~ Dave

Charts and graphs should not be too complex. Either include a handout or direct your audience's eyes to the appropriate spot. ~ Vernon

In addition to comments directed at design elements of a single slide, some were directed at multiple slides, and how to use anchors to keep from getting lost.

Use a tie back slide that lets your audience know where you are. If you are covering five items, keep coming back to your five main points to anchor where you are in the presentation. ~ Vernon

Number each slide in case someone joins late or loses connection. Comes in handy if you have handouts too. ~ Wendy

Use a visual theme that unifies the presentation despite each slide's uniqueness. Ensure that learners don't say "Is this the same presentation or a different one?" ~ Lynn

Note the different intentions of the comments. For the design of individual slides, aspects of "aesthetics" and "understanding" were emphasized; for multiple slides "anchoring" is the main theme.

Think about one of your favorite web sites, and chances are you'll find the same sensory factors – aesthetics, understanding, anchoring – at play. One example is smilebox.com. The site name, its slogan "make ecards for every occasion," and its simple design with limited text promote immediate understanding – to offer thoughtful cards that show people how much you care. The design is aesthetically pleasing, sporting an uncluttered look with interesting, appropriate visuals. The menu structure is sensible and consistent among pages, leaving no doubt as to how to navigate.

If you land on this kind of site, you have an immediate feeling that you're in good hands. Try to reproduce that feeling on each slide of your deck.

Slide content and delivery

After the category of design, the next largest general category created from my students' suggestions is content. Content is text or graphics (or audio) that conveys the full or partial meaning of your message.

Use only one clear idea per slide. ~ Sue

On the other hand, be sure to cover all the material that is on the slide. ~ Sue

Try to have the content generate thought - if they aren't clear what it means, can they take some guesses? Give people the opportunity to think through the concept. ~ Mary

Less is more - grab them with what you are saying, not what you want them to read. ~ Jen

These bullets imply that the slide content is only *part* of the message. The other part of the message is the delivery of the slide, which is the third major category that my students' comments fall into. Slide delivery includes aspects of voice (pitch, modulation), pacing and flow. My virtual training participants had a lot to say about the delivery.

Voice

Slides are NOT your presentation - they are merely support for your message which is usually delivered verbally. ~ Sue

Paraphrase, rather than reading the slide. ~ Sue

Speak emotionally to participants' right brain and back up with words to go through participant's ears to talk to their left brain. ~ Tonya

Pacing

Never find yourself clicking rapid-fire through slides while you blab a few words. If you have included too much material, skip it. ~ Lynn

Leaving a slide up too long over the web will encourage your audience to become disengaged. ~ Dave

Flow

Keep the correct sequence of information, so it flows in a logical way. ~ Veronica

To show progress, repeat slides with things checked off that have been accomplished. ~ Mary

One of the take-aways is that the content of the slide and the delivery by the facilitator must together convey the significant points. As mentioned earlier, Seth Godin suggests using both logic and emotion to drive home your message: "You can use the screen to talk emotionally to the audience's right brain (through their eyes), and your words can go through the audience's ears to talk to their left brain."

If you're an impassioned speaker or a good actor, you can do the opposite: present logical facts on the page and use your voice for emotion. But either way, know that content and emotion are conveyed by using both slide and voice.

Many of the delivery suggestions speak to proper pacing. Not too slow, not too fast. The content must flow logically, and it's important to pause to summarize or review.

Differentiating between presentations in the physical and virtual classrooms

Following these tips on content and delivery are important in both the physical and virtual classroom. So are the design suggestions. But we're now coming to some important points about how virtual PowerPoint presentations *differ* from those in the physical classroom. Consider these suggestions gleaned from my workshops.

Change the slide often. People need something to focus on. ~ Sue

For web workshops, keep a pace of approximately one slide per minute. ~ Vernon

If used in a web workshop, more slides are recommended than for face-to-face meetings. ~ Dave

With web workshops, participants can't see me - at best, they'll see a fuzzy likeness of my talking head. So the slides matter. But they don't just support your message, they contain it. ~ Sue

Yes, in the web delivery world we reverse the notion that you should have fewer slides in a PowerPoint presentation.

In virtual facilitation, PowerPoint slides serve all the purposes that they do in the physical classroom, and then some. In both settings they help to focus your audience, but that function is more critical in the web workshops. You can't just turn off the slides and redirect attention to your physical presence. But you need to use (and not use) PowerPoint wisely to ensure engagement.

The streaming video and the interaction tools mentioned during this chapter help to bridge the focus gap – as a matter of fact, it's their primary purpose. Another trick for ensuring visual attention to a specific slide is to change its layout dramatically.

The font size you choose for a PowerPoint slide in the virtual classroom must also be larger than the font used in a traditional classroom presentation. Logged into your web workshop, participants may be viewing your slides from their desktops, laptops, mobile devices, or smart phones. We recommend a font size no smaller than 28 point so that your slides will be readable on smaller devices or monitors.

My student's assertion that slides "don't just support your message, they contain it" is another important tip to remember. Notice that this differs from Godin, who suggests loading the slide with emotional graphics and letting your verbal delivery contain the content. That works wonderfully in some scenarios (like pitching a product or a solution) but not as well in a training scenario. In an instructional setting, the documentation of the content is important; the slides should contain meaningful content, either in the slide itself or the notes that accompany it.

Here's an example of a poorly designed slide intended to convey information about the amount of plastic floating in the ocean.

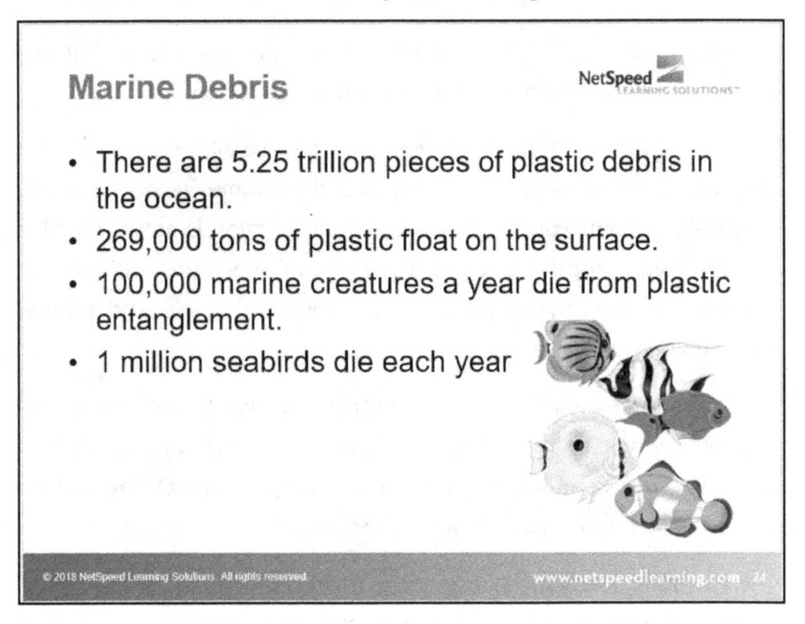

As you can see, this slide is not very effective. There is too much text with unrelated pieces of information and cheesy clip art featuring tropical fish. Sadly, slides like this one appear in the virtual classroom all too frequently.

Now look at a slide that uses an image to convey content with one key message in text.

This slide conveys the key message with an emotional punch.

Here's a summary of the powerful points we've made about PowerPoint.

- Design
  - Follow the same best practices for design elements as you would in a physical setting.
  - Design points are important for both individual slides and groups of slides.
  - Design slides with an eye for aesthetics, understanding, and anchoring.
- Content and Delivery
  - When writing content, think of your delivery. They MUST go together.
  - Content can often generate thought and discussion. Prepare for it.
  - Pacing is important. Practice it.
  - Remember to marry logic and emotion.

- Differences between the physical and virtual classroom
    - Use more slides (not fewer, as is often taught), at about a slide a minute
    - Think of slides as a focusing device, along with streaming video and interaction tools
    - The font size must be large enough that it can be read on a small mobile device or screen
    - The content must carry your message for future reference

Homework

Assignment 5-1: Attend a web workshop

Find and attend (or view a recording of) a one-hour webinar on the Internet on any topic of interest to you. Write a critique of the presenter's strengths and weakness. What did the presenter do particularly well? What suggestions would you make to improve the presentation of the webinar? Write a one-page response describing the title and content of the webinar, as well as your observations about the presenter's strengths and weaknesses.

---

[x] Godin, Seth. "Really Bad PowerPoint."
https://seths.blog/2007/01/really_bad_powe/

chapter six

repurposing from traditional to virtual

Now that you've seen how to promote interaction and collaboration, we'll explore how to convert traditional classroom training into engaging interactive web training. As more and more organizations move to online training, the ability to re-purpose or redesign an existing learning experience for the virtual classroom becomes an increasingly valuable skill.

Revisiting adult learning principles

Let's start by perusing four key principles of adult learning that Malcolm Knowles proposes in his classic book, *The Adult Learner*[xi]. These apply to any kind of learning environment.

- Adults need to be involved in the planning and evaluation of their learning
- Experience (mistakes included) provides the basis for learning activities
- Adults want to learn subjects that have immediate relevance to their work or personal lives
- Adult learning is problem-centered not content-oriented.

Knowles' emphasis on learner involvement and honoring adults' needs for a more pragmatic focus of learning activities dovetails nicely with the three basic types of learning that we identified in Chapter 3: affective, cognitive, and behavioral. When we revert to the old style of webinar and simply plod through a PowerPoint presentation, we target only the cognitive domain. By all means, give your learners content to be able to improve their performance, but strive to influence their attitudes and emotions (affective) with healthy interaction and sharing of experiences, and tie both the cognitive and affective together with the development of skills, behaviors, and performance during and after the learning session.

We need to be reminded of these important principles. Reach out to attendees to share their work experience, both ahead of time and during the session. Understand the problems they need to solve so you can revise your learning plans and make sure that the learning is applied immediately.

When you're ready to repurpose traditional classroom situations for the virtual environment, remember Knowles' principles and the ABC learning domains.

Presenting some common repurposing errors

Given that backdrop, here are some of the most common repurposing errors I see.

Eliminating experiential exercises. Webinars often do away with hands-on learning. In my mind, this is the biggest conversion mistake. We have tended to focus on cognitive objectives, like information delivery, and have given up the opportunity to engage learners in tackling real challenges.

Returning to a "talking head" presentation style. It doesn't work in the physical classroom; why would we think it's going to work in the web workshop experience?

Using polling questions as a "quiz." It's okay to use polling questions as a quiz. Yes, that's a first step. But if we're only using them to test what our participants have learned, we're losing some of the richness of the tool.

Not allowing participants to chat with each other. We too seldom encourage people to talk to each other. It's important that we keep the class sizes small enough to allow this.

Limiting questions until the end of the session (Q&A). This basically shouts the message, "Don't interrupt me, I'm presenting." Instead, direct your learners to enter questions into the Q&A pod or in a chat window of the web platform.

Eliminating personal stories and examples. The other thing we do is ignore the power of stories and examples. Due to the formality of the web conference, we tend to fall back on our PowerPoint. Not interesting. Stories are the mental hook that retains information. So, by all means, tell stories.

Reading your script or slides aloud in a sing-song or monotone voice. Nothing spells "boredom" like listening to someone lapse into slide reading. It's an easy mistake to make in a web workshop.

Using complex graphs or charts that are poorly explained. We may think we're communicating when we pop that detailed graph into the PowerPoint. In truth our participants can't figure out where to look unless we guide them, which leads me to our final repurposing error.

Not focusing participant attention (with annotation tools). You have the ability to point, highlight, and draw in most web conference software. Using these tools helps your participants follow the flow.

For the remainder of this chapter, we'll discuss some specific repurposing techniques, and give an example or two to further illustrate them.

Breaking the ice

In Chapter 4 we demonstrated a mingle that employed the tools of chat, polling and Hand Raising. Another useful exercise to break the ice is the Photo Matrix. I highly recommend getting your attendees' photos ahead of class and displaying them on a whiteboard.

Then ask each person to sign the grid next to their name. Don't forget to include yourself. I sometimes include an image in the grid that doesn't belong there; whoever identifies the mystery man or woman can type the correct name into the chat pod. A little visual humor is fun and captures people's attention. Can you find the ringer in the photo matrix above?

In addition to a fun exercise, the photos are useful to you, as instructor. I will often tape my attendees' pictures and names on my monitor as I run the workshop. It gives me a visual sense of whom I'm talking to, and is a great memory jog for associating their responses to their names.

Use your imagination with the ice-breaker. In addition to assigning a name to the face, you can ask questions, like "Tell us one thing no one else knows about you."

Conducting peer-to-peer discussions

In the Communication Styles demonstration in Chapter 4, we saw one way to repurpose peer to peer discussion: by setting up multiple chat windows. This works well for several individuals per window.

Other choices are to share a whiteboard, or use a group grid or matrix. Try using more than one option during your session to keep things interesting.

Employing role-playing or practicing

Role-playing is staged the same way in the virtual classroom as the physical one. Ask for two volunteers to unmute their phone lines. Have them play different roles – for example have one play the manager, the other the employee.

The disadvantage of virtual role-playing is the obvious one: you may only have the audio to go by. You may not be able to see the players walking, gesticulating, or grimacing. With the advent of video and web conferencing platforms such as Zoom and Skype for Business, it has become easier to put participants on camera. That is a great way to improve the experience of role-playing for everyone.

Modeling behavior

The ability to model effective behavior gives a class the chance to *experience* a desired role or skill that is performed well, instead of merely reading about it. The modeling of behavior can be repurposed from the traditional classroom to our virtual world by having the host and presenter model an effective interaction, followed by having the learners use chat to give feedback. Let's see how our fictitious character, Gerri, experiences this.

> Gerri has just heard the instructor, Cynthia, explain how you can role-play virtually. Now Cynthia goes on to talk about modeling behavior. Gerri is intrigued, but wary.
>
> *This is hard to pull off well in my classes, Gerri is thinking. The role-playing doesn't always go well because it's off the cuff. But at least the role-players aren't expected to be great actors. Modeling effective behavior is trickier, because you do need for the person to be effective.*
>
> > "To model behavior, a good idea is to bring in your host to help you out," says Cynthia. "So let's give an example of how this might work. Tim, you can come on-camera now."
>
> Tim's friendly, smiling face appears in streaming video, while Cynthia's is blacked out.
>
> *Well, there's an idea. By using the host, you get an actor who's on camera. You wouldn't get that if it were a student performing.*
>
> > "We're going to use a training module that we teach at NetSpeed Learning, called *Coaching to Redirect*, which is about how to effectively coach people on your team. Tim is going to give us an example of how *not* to give constructive feedback."

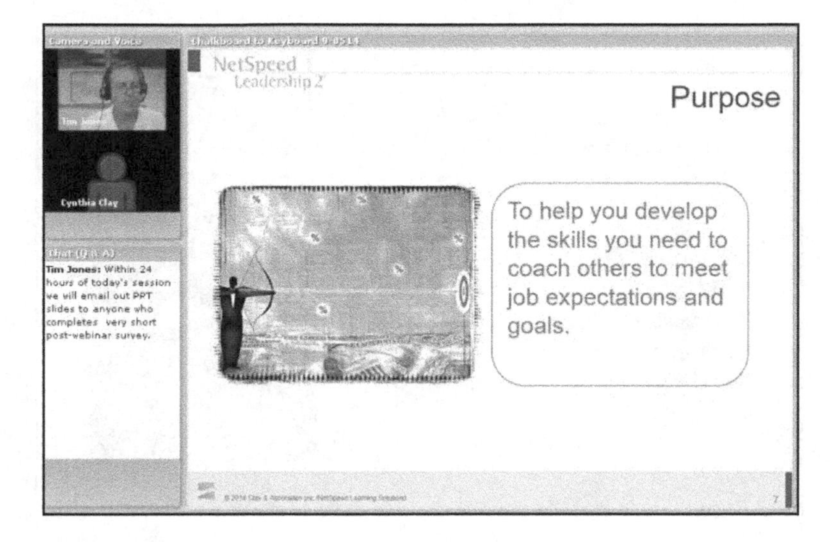

Gerri notes that Tim's expression changes from friendly to stern. His voice comes over, loud and authoritative and nasty. "Stephanie, I don't want you coming in late any more. It really irritates me and your team is upset, too."

I can hear Cynthia chuckling in the background. She's getting a kick out of this. I'm beginning to see the importance of having a host. Maybe the IT guy would help me. He can handle Q&A, and deal with technical issues. I'm starting to think I can do this.

> "Tim, thank you. Now, say we've sent Tim off to management training. He'll have the opportunity to give the same message more effectively."

A new slide appears.

NetSpeed
Leadership 2

Feedback

I don't want you coming in late anymore. It really irritates me and your team is upset too.

Stephanie, I noticed that when you don't communicate about being late to your team, they can't prepare for your absence. As a result we don't have adequate coverage at the counter.

© 2014 Clay & Associates Inc./NetSpeed Learning Solutions 34

Tim has apparently mellowed. "Stephanie, I noticed that when you don't communicate about being late to your team, they can't prepare for your absence. As a result, we don't have adequate coverage at the counter. This causes a longer wait time for our customers."

> "You can bring that host voice in to represent another voice on the call. And it just wakes people up. It's the radio show phenomenon, the banter."

My IT guy will love this. I think.

> "If I were teaching the module, I could then show a six-step model. I can model each of those steps."

Up goes another slide.

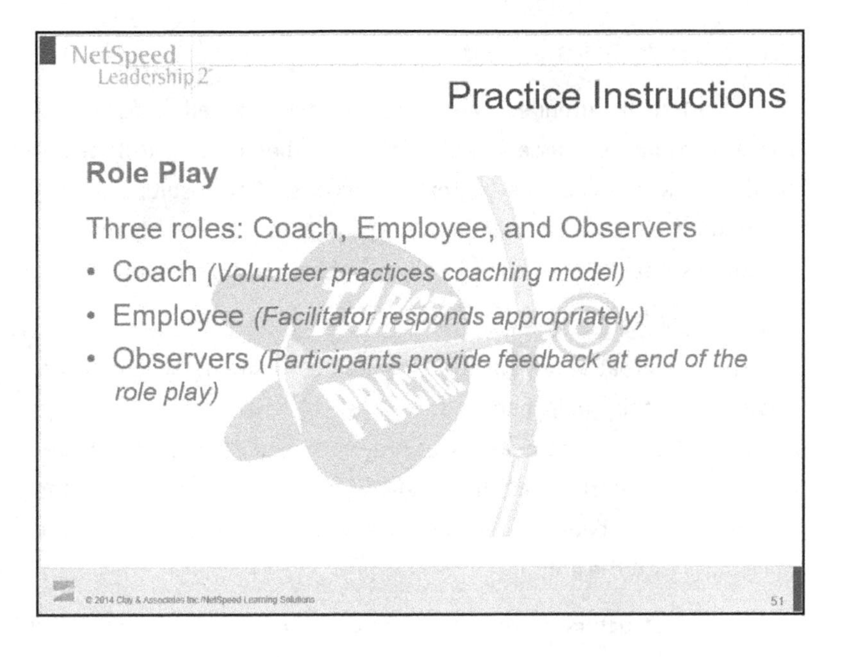

"Then I can bring you into a practice session where I can divide us into the coach, employee, and observer roles. I might send you into a break-out room in groups of three and make sure you're rotating the roles to give everyone an opportunity."

I'd like to see that.

"Or do it in the full room as a fishbowl and get practice with all of you being observers."

I'm still concerned about getting a host who can pull off both the tech role and the role-playing. I'll use the Q&A pod to ask Cynthia about it.

Cynthia responds immediately.

"Sometimes I call ahead of time to get a plant in the audience who is willing to practice the model as coach. You can then invite anyone to volunteer, and keep your shill in your back pocket as needed. They will volunteer if others won't."

Employing self-assessments

Occasionally attendees are asked to fill out self-assessments. These can include assessments of skills, behaviors, attitudes – anything that matches the context of the class. The results of a self-assessment exercise is sometimes surprising to the participant, and sometimes not. Perhaps the more eye-opening aspect is what it tells them about themselves in relationship to others.

That's a great argument for using self-assessments in the virtual classroom – they help people talk to each other. I send a link to complete the assessment survey as pre-work. You'll remember that in Chapter 4, Gerri and her fellow learners completed the Communication Styles assessment in class; then I used polling to share the results with the group.

Some instructors make it a rule not to send handouts, but I like them because I'm looking for as much variety as possible. I think their reticence is related to why we never send out PowerPoint slides in advance – you don't want students to feel like they know what the session is all about, which is a license to multi-task. From my experience, as long as you can keep your learners so engaged that they're afraid to miss something, handouts are fine.

Charting ideas on a flipchart

In a physical room we often use a flipchart to capture ideas. The virtual equivalent is the whiteboard. Have your group unmute their phone lines to start the discussion, then you, your host, or a volunteer can enter the ideas into the whiteboard as they start to flow. Of course you could use a chat window for that too, but remember that the whiteboard has the advantage of using both text and drawings.

Asking questions and checking the pulse

I've given several examples where I ask questions from learners. Some used the Q&A pod or chat windows. In Adobe Connect, you can have multiple chat windows open for small groups.

For open-ended questions, pose an interesting question and ask for input. Use polling and offer multiple choice responses. Allow learners to see the results, then chat. While reading the chat responses, ask for additional clarification.

You saw an example of another way to solicit answers in the Communication Styles simulation two chapters back. I asked the Anchors to raise their virtual hand, then unmute and share answers to questions.

For closed questions, I like to use the status icons on the Adobe Connect platform. These are sometimes called emoticons. In the beginning of the session I may ask people to use the raise hand icon to signal, "I'm ready to learn three new techniques." Or I'll ask for examples: "Does anyone have an example of a time when...?" There's also a status icon with a green check and red "X" for Agree and Disagree, as well as an Applause symbol (clapping hands). In our *Virtual Facilitator Trainer Certification* series, everyone delivers a web workshop on the final day. It's great to "see" that virtual applause when you're done.

Some platforms offer a Thumbs Up/Down icon. I might use that if I want to check the pulse of the class regarding pace or content. Another trick we've seen is to present a closed question with any of these icons, perhaps following a poll. Then call on people by name to expand on their answers. There are a lot of creative ways you can use the web conference platform tools. Explore them.

Here's a recap.

Traditional Classroom	Virtual Classroom	Chat	Q&A	Polls	Status Icons	Audio
Asking for questions from learners	Question & Answer Chat: Use separate chat or specific Q & A pod for questions from learners.	√	√			√
Asking open-ended questions	• Chat: Pose interesting question and ask for input • Polling: Offer multiple choice responses; allow learners to see results. • Then Chat: Ask for additional clarification.	√		√		√
Asking closed questions	• Icons: Use raised hands or thumbs up to indicate agreement; thumbs down to show disagreement. • Polling: Present closed question with Yes/No response; show results. • Then Call Out: Call on participants by name to explain their response.			√	√	√
Checking the pulse of the class	Icons: Ask for thumbs up or down regarding pace or content.				√	

Presenting a video

In the classroom, we often present a video and then discuss its message. In the virtual class, you might send a link out before the web session, or you might choose to upload a video into your web conferencing platform. Make sure it's not too long, perhaps one to two minutes. In Adobe Connect and WebEx you can upload a video and show it right in the meeting. It will play over your participants' speakers. Some web conference platforms don't allow you to play video through the system. In those cases, an option is to use application sharing: simply run the video from your desktop and share it so that everyone views it. (You may have to hold your phone handset or headset to your computer speakers to play the audio, so test this ahead of time.) Whatever option you decide to use, practice your solution before the actual class with other people logged into your dry run or rehearsal. There may be difficulties in some platforms with streaming a video so that it plays at the same speed for every participant.

You can use video to add a new face and voice to the session or to present a simulation or role-play for participants to observe. In our web workshop *Creating an Inspiring Work Culture*, we link to a five-minute video produced by Playing for Change[xii] featuring musicians from around the world collaborating on the song *Stand by Me*. It's very inspiring and a wonderful way to affect the attitudes of the managers in attendance.

Presenting with PowerPoint slides

When repurposing your slide presentations for a web conference, don't forget the points we made in previous chapter, especially the ones in the section called *Differentiating between presentations in the physical and virtual classrooms*. Use *more* slides via web conference (with, of course, interesting graphics) because we're using slides to manage attention. Don't spend any more than one minute on each slide.

Another rule of thumb I use is that there should be no more than five minutes between participant interaction and collaboration. They should get to do something, so they're not just listening to a talking head.

Checking for understanding

To check for understanding among your learners, ask them questions and have them respond with a status icon. You might ask them to recall a time when they received constructive feedback from their boss, for example. Have them use the raise hand or green check icon if they recall how it felt to get that feedback. Then use chat and ask for examples of the feedback received and how it felt to receive it. Call on someone who is actively participating to unmute their phone line and share their experience. In this way, your learners help each other.

Coaching learners

To convert coaching exercises from the traditional classroom, I usually present a sample statement, then ask whether the statement is effective or ineffective. I then open chat and ask how participants might improve the example.

Coaching is another one of those skills that takes some practice. Perhaps it's a good time for another demonstration.

Cynthia is speaking.

> "When you deliver your web workshops, you can also coach learners by presenting a statement, asking for feedback, and discussing the feedback. All the while you have to reinforce

the 'correct' answer, if there is one, and clarify points that may be confusing.

"So for example, if I'm teaching NetSpeed Leadership's[xiii] *Appraising Performance* module, we might set up the Hall of Shame and the Hall of Fame, and let you know that a comment belongs in the Hall of Shame if it's judging, vague, subjective, or not motivating. Conversely, a comment belongs in the Hall of Fame if it's descriptive, specific, objective, and developmental."

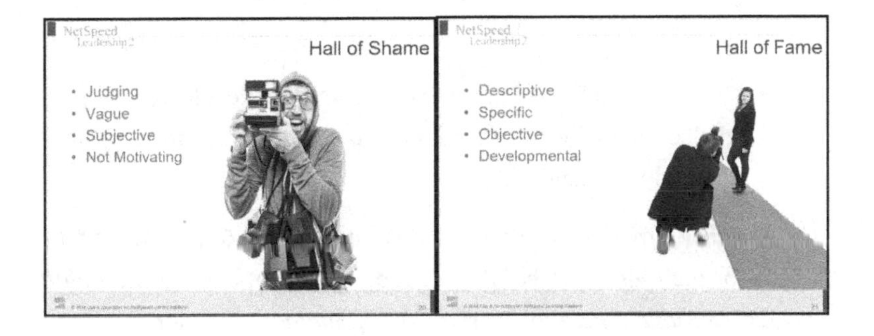

"I'll then give you a chance to respond to some of those by offering a statement and have you decide if that statement belongs to the Hall of Shame or the Hall of Fame. Here's our very first statement that might appear in a performance appraisal."

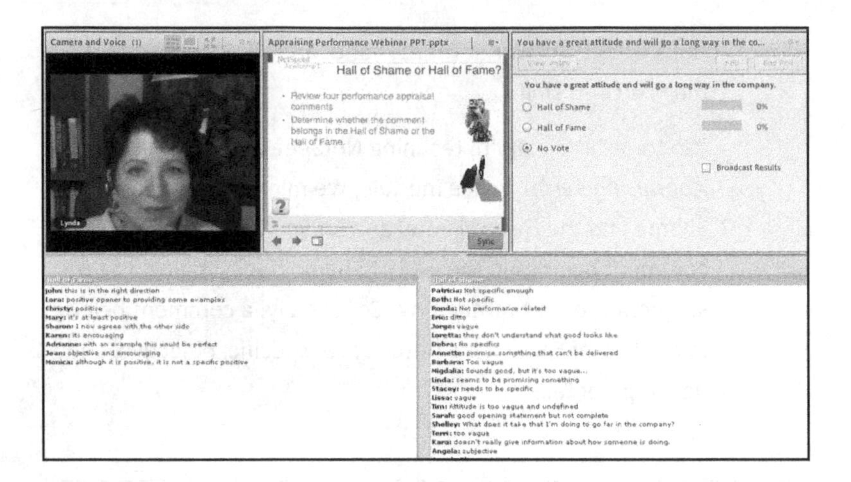

> "It says, 'You have a great attitude and will go a long way in the company.' Where would you place that? Use the poll question to respond."

Hmmm, Gerri thinks. It's a positive statement. But it doesn't fit any of the four Fame points. It's too vague. Not exactly shameful, but it probably belongs in the Hall of Shame. I wonder what the others think.

The poll results are still blank on Gerri's screen.

"I'm watching those poll results come in. I don't want to say if it's right or wrong. Okay, it looks like everyone has responded, so I'll pull these up."

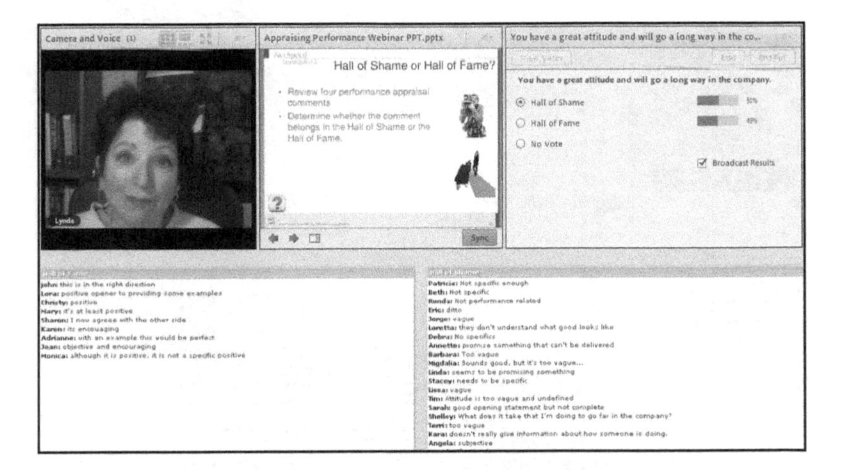

"As you can see, the results were very close. 51% placed it in the Hall of Shame, and 49% in the Hall of Fame. Now feel free to chat. If you believe it should land in the Hall of Fame, chat in the left box. If you think it belongs in the Hall of Shame, chat in the right box."

Here goes, thinks Gerri. She types in "Too vague" in the Shame section. As she reads the other results, she finds a lot of agreement.

As the chat results come in, Cynthia does a running commentary.

"I'm looking at the Hall of Fame. Christy says it's positive; Karen says encouraging. Others echo that."

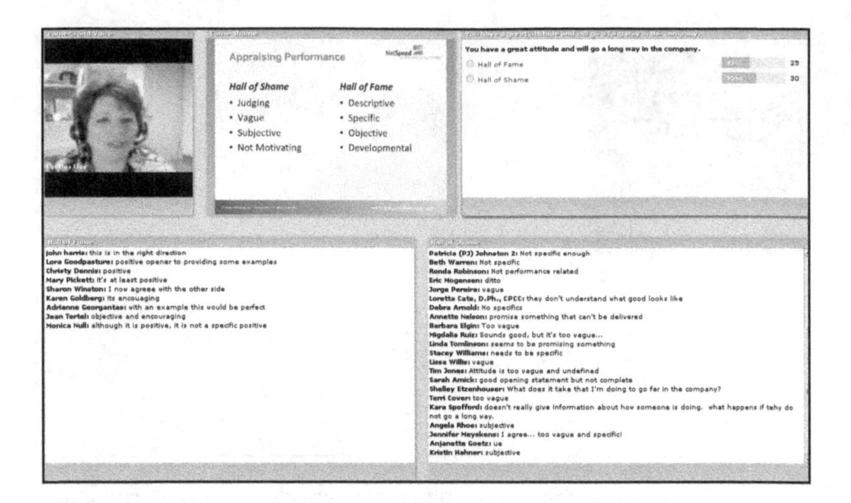

"Over in the Shame camp, Alejandra doesn't think it reinforces the strengths. I see terms like not truly a compliment, judgmental. Gerri thinks it doesn't give useful feedback. Gerri, why don't you unmute your phone and tell us about that."

"Hi, Cynthia. It was an upbeat message, but it's not constructive. So because it's vague I put it into the Hall of Shame."

"Yes. This is the major takeaway for this one. You could argue that a pat on the head is motivating, which is a good thing, but the vagueness places it into the Hall of Shame.

"Now in our regular *Performance Appraisal* class I would ask more questions about why you selected what you selected, and make my learning points. We could discuss how we could change the statement to make it more descriptive and specific. But this should give you some taste of how to coach others through a subject using polling, chat, and conversation in this way."

She's right. It's amazing how much is going on. We're being engaged, we're thinking about what to say, making judgments, sharing thoughts, hearing what our peers have to say, all happening in real time. We may not be chatting voice to voice, but we're actually participating in an evaluative discussion. I can't wait to use this!

Homework

Assignment 6-1: Repurposing three classroom exercises

Read the repurposing samples in Appendix C. Then repurpose each of the three exercises, using your own ideas. Appendix D will help you select appropriate interaction tools for your repurposing exercises.

[xi] Knowles, Malcolm, Elwood F. Holton III, and Richard A. Swanson. The Adult Learner, Sixth Edition: The Definitive Classic in Adult Education and Human Resource Development. London: Elsevier, Inc., 2005.

[xii] The video, produced by "Playing for Change," features musicians from around the world collaborating on the song *Stand by Me*. The URL is www.playingforchange.com

[xiii] NetSpeed Leadership is a facilitated classroom and online training program for supervisors and managers. It emphasizes practical theory using tools that can be applied immediately in the workplace. http://www.netspeedlearning.com/leadership/

chapter seven

learning transfer

I love this quote from Allison Rossett of San Diego State University: "Transfer of learning is the application of skills and knowledge learned in one context to the context that matters."[xiv]

It's so true, isn't it? A linebacker on the practice field is having skills and knowledge being pounded into him so that he can perform in the context that matters: the game on Sunday. A new hire in a manufacturing plant shadows an experienced worker so she can perform when it's her turn to step up to the assembly line.

In both of those examples, the learning contexts are just a step removed from the context that matters. How many steps removed, would you say, is the physical classroom? What about the virtual classroom? How well do you think we are doing in the transfer of learning?

Here's one person's assessment: "Although transfer has been almost universally recognized as fundamental to all learning and must therefore occur all of the time, the history of research findings on transfer suggests it seldom occurs in instructional settings." The quotation is from Robert E. Haskell's book *Transfer of Learning*[xv].

As instructors, we need to keep firmly in mind that the instructional setting is not the main event. Our great challenge is to make the transfer of learning to the context that matters, whether that is a social setting, an assembly line, or an executive boardroom.

Converging to the context that matters

The traditional model of learning assumes an instructor talks (or writes) about skills, knowledge and behavior to students who attempt to assimilate the instruction. This is also the model most people have in mind when they think about webinars.

Dr. Henry A. Giroux (McMaster University) noted, "Where I grew up, learning was a collective activity. But when I got to school and tried to share learning with other students, that was called cheating."[xvi] Hilarious, but true. In the traditional classroom setting, the teacher is the center of the educational universe, the sole conduit to knowledge. Conversation with other students is considered a distraction, not a contribution, to the task at hand.

In adult education, we may need to unlearn what we learned in school. The skills, knowledge, behavior, and attitudes that comprise the learning experience should be gained from as many sources as possible. The environments in which the learner learns should be varied, and, ideally, should simulate the "context that matters" as closely as possible. *Convergence* occurs when learning is accessed and reinforced from multiple sources.

One convergence strategy is to develop collaborative communities. A tremendous benefit takes place when a community of learners shares what they are learning and builds on each other's contributions. These are often called "communities of practice." Giroux's neighborhood is an example. The classroom can be another. A simulation of the actual setting (e.g. the practice field, the assembly line, or a team-building exercise) is yet another.

The context that matters is the actual work environment. This is often the last stage of the learning experience. Think of a brain surgeon who has studied, researched, practiced in low-risk situations, and interned in a hospital before she ever applies a scalpel to your skull.

Other convergence strategies require individual effort. Learners can access knowledge bases, libraries, and online performance support systems. For example, in all of the NetSpeed Leadership courses, learners are expected to complete an online performance support tool (NetSpeed On The Job[xvii]) within a week of the instructor-led classroom experience. In most cases, they do this from their desktops or a mobile device when they are back on the job. Of course, once this information is transferred to the context that matters (that is, once it leads to behavior, knowledge or attitude that is exhibited in their place of work), the individual can also share it with others.

Putting the learner at the center

We talked earlier about the old-school image of an instructor as the center of learning. Marc Rosenberg's[xviii] work on learning transfer suggests that we shatter that image and instead make the *learner* the center of learning. As we design learning experiences and we surround the learner with resources, we assist this convergence process. Ultimately, learners are responsible for their learning and improved performance but they are better able to integrate what we are teaching them in the classroom with what they are doing on the job.

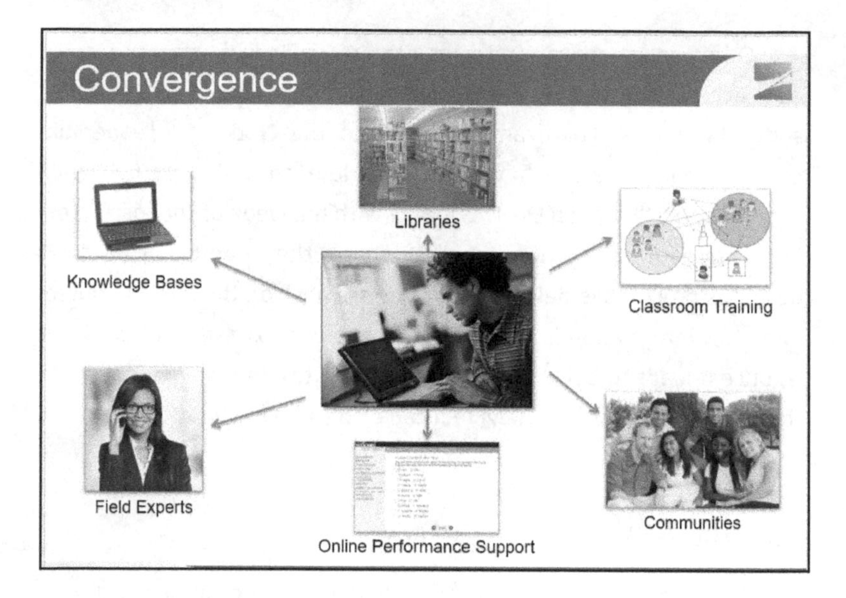

Learners can learn in the web conference platform or face to face. They can share best practices with communities like Internet forums and professional organizations. They need performance support tools like short videos, job aids, and help desks made available to them. They need access to experts and mentors, knowledge bases and libraries. Company programs like on-the-job training and job shadowing can help. This interconnecting of resources and integration of input is what makes learning stick.

Convergence and the facilitator

Perhaps you're still thinking that your role as a facilitator stops with the virtual classroom, and that accessing other resources is the sole responsibility of the learner.

But we can do more than that – we can integrate many convergence techniques into the web workshop itself. This is a good time to check in on Gerri. She's been listening to Cynthia's explanation of convergence, and may be a bit overwhelmed, since this is all new to her.

> *This is not what I was expecting. I can't see how to incorporate convergence into my physical classroom setting, much less my virtual one. I'm still not sure I even know what convergence means. Am I the only one who doesn't get it?*

The streaming video display of Cynthia continues confidently on.

> "Let's take this one step at a time. The first step is to do our homework before the web workshop begins. We should gather information to help us know the environment of the participants."

A new slide is displayed.

> "Let's look at several good questions adapted from those posed by the authors of *Principles of Human Resource Development*.[xix]"

Good Questions

- How are skills, knowledge, and attitudes being used on the job?
- What barriers prevent learning from being transferred?
- What role do managers and supervisors play in transferring and reinforcing learning on the job?
- What should be done before, during, and after implementing a course to enhance learning transfer?

> "The first three bullet points might uncover challenges. Hopefully they may also provide opportunities."

"Hmmm. These look familiar," thinks Gerri.

> "Unmute your phones and let's discuss these points. Tell me what you think."

The first voice is a male who Gerri recognizes as Fred. "The first three bullets are similar to the ones we discussed when we talked about learning objectives."

He's right. That's where I've seen them.

"It's starting to make more sense to me," Fred continues. "It's good for us to know as much as possible about the context that matters to the learners. Not just so we can design better exercises, but because having students practice the exercises brings them as close as we can get them to the real world. It's simple, but I guess I've never looked at it in this way before. I've always just focused on what content I need to package a presentation."

"This is Carol. I think one of the challenges in the workplace is that students – and here I include myself – are used to going to one place to gather knowledge. It's hard to understand that training involves all the other elements."

> "That's one of the barriers to adoption," Cynthia replies. "We've trained people to think that if they go put their butt into that classroom seat for a day, or they go to three webinar classes for ninety minutes, they've fulfilled their responsibilities to the learning process. Check it off your list and return to work."

Tonya speaks up. "We need to get decision makers to see it."

But this is all theory. What does it have to do with us? I'm not into politics.

There's a flurry of voices now. Someone adds, "But sometimes the managers are the ones who are putting up the barrier."

"One of our managers requested that we simply repurpose an existing class into a webinar. I know now that won't work. But the mindset is still there."

> Cynthia again. "Managers are critical players; if you leave them out, you lose a rich opportunity to get support within the system. And you're right; management often has the same mindset as our students. Which is problematic, since the return on a company's training investment is hard to prove if all we're doing is attracting people to the classroom experience."

Boy, this is all so true. But I thought she was going to talk about how we can incorporate convergence...

> "Here are some more useful questions. These are starting points to spur us into thinking beyond just the classroom training event."

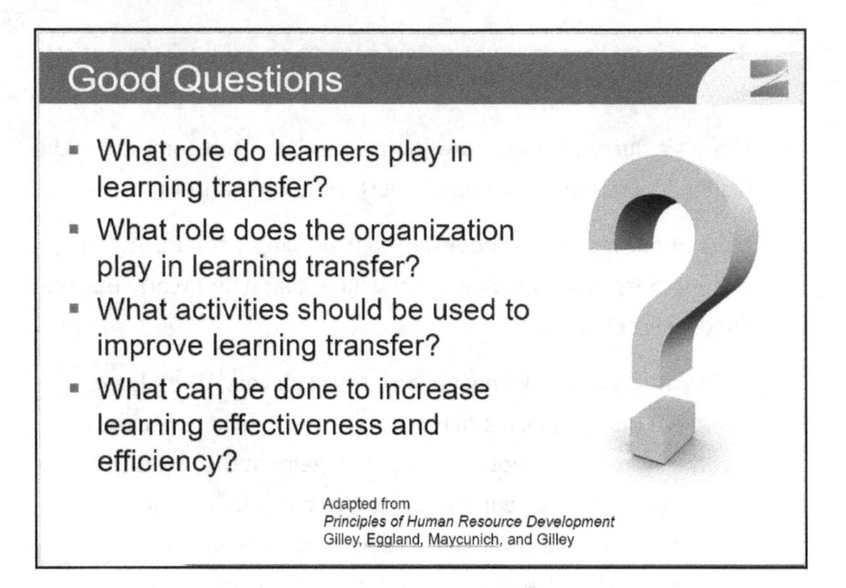

I'd better speak up. "Cynthia, this is Gerri. I have a question."

"Gerri. Great to hear your voice. What is it?"

"Even if we know about convergence, and let students know that convergence is critical to learning transfer, at the end of the day we're only responsible for one piece of convergence, the classroom training. Isn't that right? Or am I missing something?"

"Gerri, that's a great question. We've been leading up to it, and now it's time to address it. It's a good segue into your homework."

Homework?

Homework, indeed

The old-school model of education is trainer-centric. It incorporates these characteristics:

- Formal learning
- Publishing model
- Explicit knowledge
- Instructor-led
- Scheduled
- Defined curriculum
- Classroom-paced

What does homework look like in this model? That's easy. It asks for a regurgitation of explicit knowledge learned in the classroom from the instructor. It is handed in to the instructor and seldom shared with the class. It is graded as correct or incorrect.

In the *VFTC* course, learners are expected to post assignments online. Indoctrinated by the old-school education model, learners often expect the instructor to evaluate every post. It takes time for them to realize that they are posting their observations for their peers, and not their instructor.

By contrast, the new model is learner-centric. Here's how it lines up with the old model.

Traditional, trainer-centric model	New, learner-centric model
• Formal learning	• Informal learning
• Publishing model	• Sharing model
• Explicit knowledge	• Tacit knowledge
• Instructor-led	• Peer-to-peer
• Scheduled	• Just-in-time
• Defined curriculum	• Initiated by the learner
• Classroom-paced	• Self-paced

If you were to visualize homework in the new model, it would most likely represent a collaboration among students and a sharing of information (including feedback) that was gathered from many sources, as well as from the experience of the student themselves. In this learner-to-learner model, every student is a potential trainer or teacher. The key to a robust learning experience is a sense of community that may take time to develop.

For this model to work, the instructor must release control. I mentioned in Chapter 1 that one of assignments in the *VFTC* course that I give is to critique a webinar. Participants find, pick, observe, use a critique form to think about how it went, then report back on what they saw. This is an example of releasing control: send learners out to do research and report back.

As I'm designing exercises and assignments, I keep my eye on having learners collaborate together and get the sense that they're part of a community of learning. It's that frame that makes the richest learning experience possible for your learners.

This emphasis on releasing control and promoting community does bring some logistical challenges. One way to manage pre- and post-class assignments in this new paradigm is to offer online resources that support social learning, such as class blogs and discussion threads. As a matter of fact, that's what we've done at NetSpeed Learning. Our participants utilize our *NetSpeed Fast Tracks*[xx] social learning platform as a home base for peer-to-peer communication, content sharing, and asynchronous homework (not to mention session introduction and class preparation).

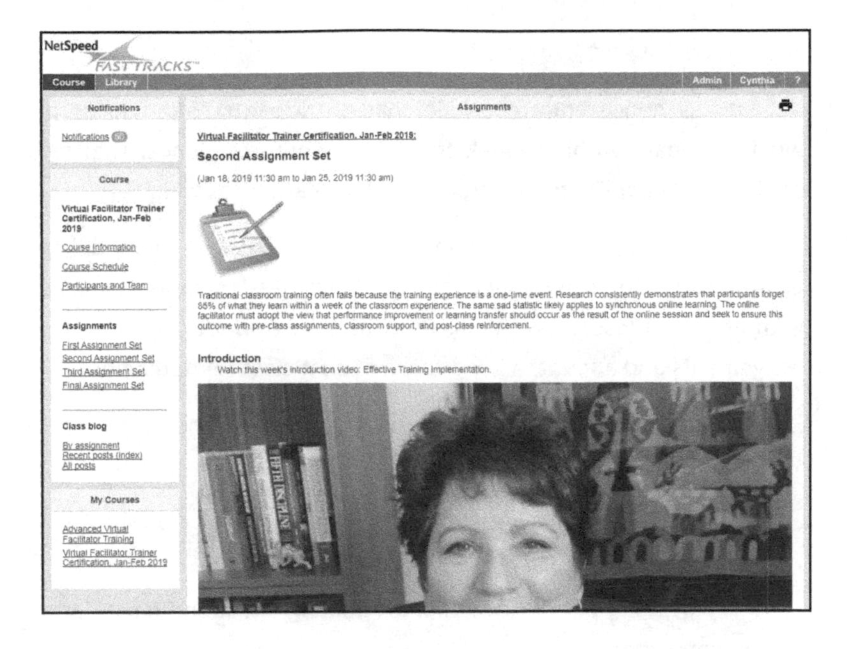

For example, one of the homework exercises at the end of this chapter is "Assess the effectiveness of a virtual training program." If this question were asked in one of our virtual classes, participants might create a class blog post, a short video, an audio podcast, or an interview. In addition to posting homework, they would be encouraged to view and comment on their peers' offerings. Collaboration is encouraged by allowing class blog posts and articles to be co-authored.

NetSpeed Note

NetSpeed Fast Tracks is a virtual learning platform that incorporates social media tools like videos, podcasts, and blogs into an integrated social learning system. NetSpeed Fast Tracks can be licensed by your organization so that your internal classes are branded and secure.

To learn more about NetSpeed Fast Tracks, refer to Appendix E.

Applying learning quickly

Studies consistently conclude that participants forget most of what they learn within a week of the classroom experience. That sad statistic likely applies to synchronous online learning as well.

We've talked about the importance of creating new online environments to manage pre- and post-class assignments. Coupled with utilizing these tools, we have to be smart about the timing of assignments and about the importance of post-class reinforcement.

According to Robert Brinkerhoff[xxi], training programs are successful when trainees:

- Apply the learning soon after training
- Have a realistic expectation of training and identified at least one application
- Are prepared and supported by the manager
- Receive incentives, rewards, and encouragement
- Engage in training close to a pressing need
- Are given tools and resources to apply learning on the job

Let's discuss each of these points more closely, and highlight some key actions that we as instructors can take to increase the chances of success.

Apply the learning soon after training

Trainees are eager to learn when they must tackle situations that require the knowledge and skill being introduced. Learners are most open to information, coaching, and support in these "teachable moments." If their work requires immediate application of the skills they have just learned in an online class, they will be more likely to change their past practices or behaviors to integrate the newly learned skill.

The key is to ensure that learners solve real-world problems in the online class. Use action planning activities at the end of each session to help them commit to applying skills on the job.

Have a realistic expectation of training and identify at least one application

Participants who use training as a day off from work get little value from the experience. On the other hand, learners who understand the goals of the class are often more receptive to the content. Participants who have a specific problem or issue in mind are more likely to share their perspectives and seek their peers' insights.

Communicate the course objectives, agenda, or plan to the participants. Help learners identify specific on-the-job challenges prior to the session. If appropriate, have the participant complete a learning agreement in which they write down their challenges and personal goals. During the session, encourage participants to identify techniques and skills that they will apply to their challenges.

Be prepared and supported by the manager

Training research consistently demonstrates that managerial support is critical to the success of any training program. Participants who have their managers' active support are more likely to attend to the content, are less likely to be pulled from class for other projects, and are often more motivated to apply the content.

A key here is to communicate the course objectives, agenda, or plan to the participants' managers. Encourage managers to discuss their goals for the participant who is attending the training session. If appropriate, have the manager review the participant's learning agreement and add any additional goals or challenges. After the training session, ask the managers to meet with participants again to review their goals, challenges, and progress. Encourage managers to use these conversations as coaching opportunities.

Receive incentives, rewards, and encouragement

A trainer or facilitator may use incentives during the session (for example, rewarding students who actively participate in chats, polls, and group discussions) and may also reward participants who actively apply what they've learned (for example, writing up case studies to demonstrate on-the-job application of skills).

However, for the most successful training implementations, rewards and encouragement also come from the manager (for example, documenting class attendance and results for the participant's performance evaluation or requesting that the participant share what they learned at team meetings). In addition, incentives may be embedded in the training system (for example, offering continuing education credits, course certificates, badges, and letters of recognition from higher level management).

Facilitators can help by developing a plan for systematic rewards from the trainer, the manager, and the organization. Ensure that participation in training receives visible recognition. Report on successful case studies in newsletters and other organizational publications.

Engage in training close to a pressing need

Avoid training people for roles or challenges that they may face months or years later. Instead, make it possible for people to access training or skill development when they most need it. This is an essential difference between education (often more focused on meeting cognitive objectives) and training (often more focused on practical, behavioral objectives). Like good coaching, training that occurs close to the student's need is perceived as relevant and useful.

As instructor, it's your job to make sure that the right people are attending your classes at the right time. Ask students to assess their need for training and demonstrate that they are attending programs at the right time in their own development. Don't use training as a perk or reward for good performance.

Get tools and resources to apply learning on the job

One of the systemic obstacles that trainees face is the lack of ongoing support when they return to their familiar jobs and routines. Make sure that trainees may call on coaches or peers when needed. Ensure that students have access to tools and online resources. Consider providing mobile reinforcement with scheduled text messages. Provide microlearning elements that can be accessed easily online after a virtual classroom session.

One way to help students with this is to plan follow-up activities, including peer-to-peer discussions, published resources (for example, searchable content online), access to an available coach for problem-solving, and progress review sessions.

The new blend

In this chapter we've quoted a variety of researchers, educators and experts on learning theory, including Allison Rossett, Robert E. Haskell, Henry A. Giroux, Marc Rosenberg, and Robert Brinkerhoff.

Given a better appreciation of the way adults think, the new model of education that is emerging, and the capabilities of new technology to support the classroom, a more complete picture of how to assure learning transfer is, I hope, being painted in your imagination.

It's a picture of what we at NetSpeed Learning Solutions call the "new blend." This is an amalgam of synchronous and asynchronous learning opportunities and resources that span the spectrum from formal to informal, and from virtual to traditional.

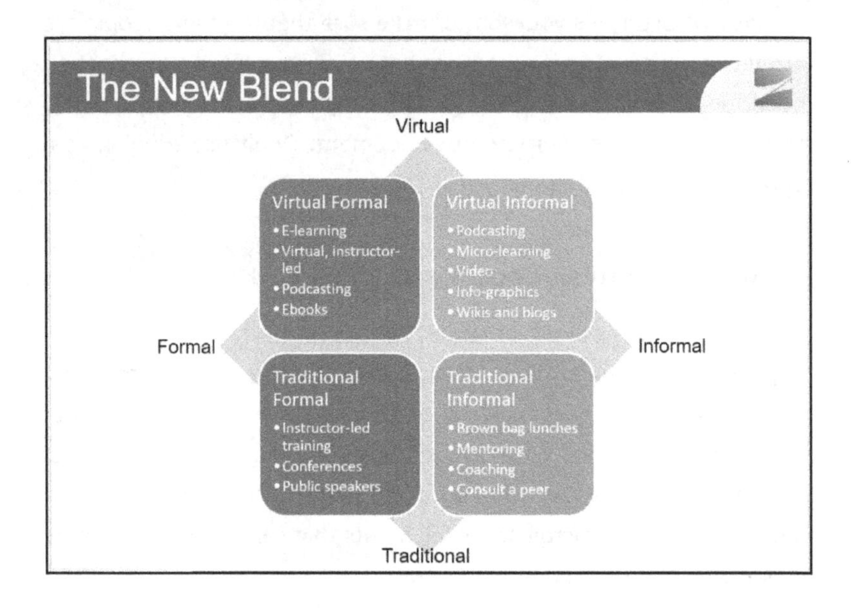

In the past, you might have mixed face-to-face classroom instruction (in the lower left quadrant) with eLearning (in the upper left). The idea may have been that you can take a PowerPoint slide deck, record an audio track, and simply stick it online.

In additional to those formal settings, the new blend adds interaction and engagement. It extends beyond the physical and virtual classroom to include informal learning, using both traditional concepts like brown bag lunch seminars and mentoring, as well as virtual experiences such as blogs, microlearning, and other social media tools – all opportunities to share with each other.

Changing the profile of learners and facilitators

There are more resources and tools at our disposal than ever before; the new blend pulls the best of all these techniques to meet the needs of your learners.

The profile of your learners is changing as well. They must now assume responsibility for their learning. They're expected to complete assignments outside classroom as well as participate in classroom.

It's always ideal to have learners meet face to face, but they can't always do that. Virtual learning has to encourage the building of relationships in other ways. Yes, the new blend includes taking a self-paced eLearning application all alone at a computer. But we know that learners will get the most value by sharing expertise and experience with each other. Consequently, they'll be expected to create and post content, as well as enter into a facilitated web conference session at a designated point in the process.

What does this mean for you, as facilitator? There are obvious challenges. A somewhat typical 8-hour, 3-day program won't work for participants sitting at a computer. At NetSpeed Learning Solutions, our approach is to create a series of web workshop experiences. These are supported by asynchronous microlearning elements that occur at our NetSpeed Fast Tracks site, which gives a combination of self-paced work and designated time to do problem solving with peers. Learners can interact both inside and outside of the classroom.

Remember: as a trainer, the goal is to develop social learning and collaboration. The story is no longer about you onstage delivering a great training experience. You still have to design and structure the learning experience, but your role has evolved into making sure people have collaborative online tools that ensure an interactive experience. You have to blend synchronous activities (showing up for a web conference event) with asynchronous activities (blogging and commenting, or co-authoring an article or podcast). There's never just passive observing in either of those places – interaction is the name of the game.

And you want to make sure that the activities are richly framed with stories and cases, because all the research[xxii] shows that if you want your learners to retain your message, it's got to come alive for them, it's got to be personal, they've got to hear your stories and tell their real life stories. All this has to be offered in a way that makes them think "Wow, I want to solve that problem."

The immediate, real-life problem solving urgency has got to be part of it or they won't apply what you're trying to deliver. As a trainer, when you guide learners to solve their on-the-job problems, create activities that allow you to release control of the content to give them a chance to come to the table with their stories and needs.

Homework

Assignment 7-1: Assess the effectiveness of a virtual training program

Identify a virtual training program and use the criteria listed in Appendix F to assess its effectiveness.

Assignment 7-2: Learning Transfer for Your Web Workshop Identify the specific actions you plan to take before, during and after your 20-minute web workshop to ensure that learning transfer and performance improvement occur.

[xiv] Rossett, Allison. "Never Worry About Transfer Again: 5 Strategies for Converging Learning and **Work**", from a course delivered at San Diego State University, copyright 2006. http://edweb.sdsu.edu/Courses/ED795A/key1/never_worry_795a.ppt

[xv] Haskell, Robert E. Transfer of Learning. 2001.

[xvi] Giroux, Henry A. Teachers as Intellectuals: Toward a Critical Pedagogy of Learning, Bergin & Garvey Publishers, Inc., 1988

[xvii] NetSpeed On the Job

[xviii] Rosenberg, Marc. *Beyond E-Learning: Approaches and Technology to Enhance Organizational Knowledge, Learning, and Performance*, copyright 2006, John Wiley & Sons, Inc.

[xix] Gilley, Jerry, et al. Principles of Human Resource Development. Cambridge, MA: Perseus Books Group, 2002.

[xx] NetSpeed Fast Tracks is a customizable, online learning platform that allows participants to engage in self-paced learning activities in a virtual, collaborative learning environment. http://www.netspeedlearning.com/fasttracks/

[xxi] Brinkerhoff, Robert O. Telling Training's Story: Evaluation Made Simple, Credible, and Effective. San Francisco: Berrett-Koehler Publishers, Inc., 2006.

[xxii] Schank, Roger

chapter eight

technology trauma

In the very first chapter, I shared horror stories about technology problems encountered while giving web workshops. They may have seemed a bit frightening, especially if you're not particularly technologically savvy. But sometimes a little alarm is a good thing. It will prompt you do everything in your control to reduce technological irritants.

The purpose of this chapter is to guide you in how to anticipate technology issues, work to prevent them from happening, and, when all else fails, make sure you know how to defuse them when they occur.

Making it look easy

Remember the first time that Gerri was exposed to an interactive, engaging web workshop dialog in Chapter 2? The positive experience changed her attitude toward web training. But consider what would have happened if, in that example, Cynthia had experienced technical issues that were beyond her control. The disruption could easily have reinforced Gerri's fear of technology and pessimistic perspective toward virtual facilitation.

So let's repeat our original example, this time presenting some real-life technical challenges and how an experienced facilitator would handle them. We'll pick up from the introduction of the Housekeeping slide.

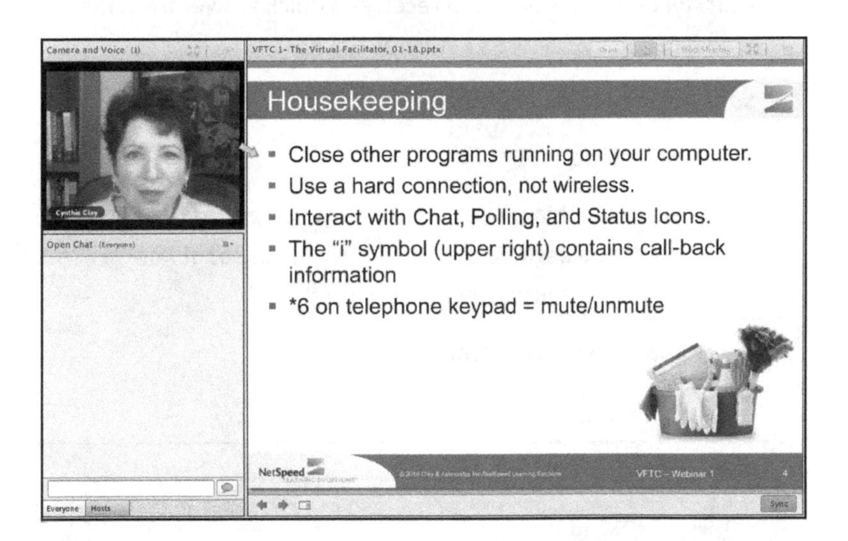

Cynthia says, "These housekeeping rules will make the experience more positive not only for you as the eventual presenter, but also for the participants. For best results today, I'd like you to close out other programs that might be running on your desktop. Web conference platforms use lots of bandwidth and you may find that having other programs open interferes with the performance of Adobe Connect."

Hmm. I was hoping to get my project done during this webinar. I won't be able to do that if I close the program. I wonder if we can ask questions during the session. Well, I don't want to bring that up now because that will slow everyone one down and I could look foolish.

"During this session if you have questions, send them right to me or our host and we'll help you. Just use the Open Chat box below my video. This is a great way for us to interact with you without distracting everybody. Only we can see it."

Oh. Okay.

Gerri types in a question and receives a quick answer from Tim, who had been introduced earlier as the host.

So the host is helping out. Smart.

Cynthia continues:

> "One of the things I like to do is help you get engaged in the tools. In the beginning we don't talk much about *how* to use them, we just jump right in and use them."

A poll appears with four options.

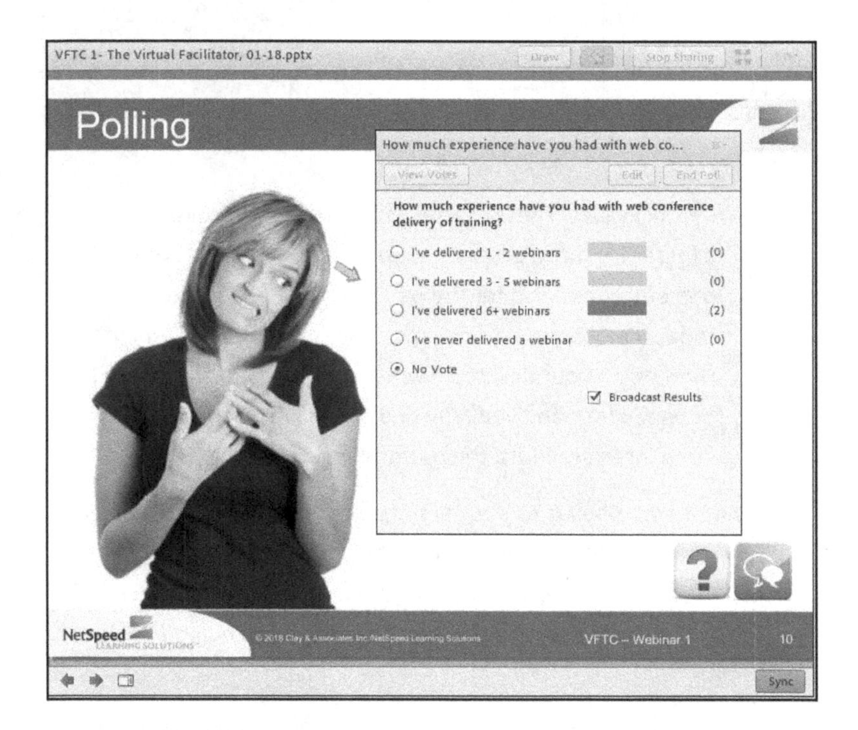

"Cynthia, this is Mary. Do I have to press a submit button anywhere for you to see my poll response?

> "No, clicking the button to the left of your preferred answer sends your poll response. That's all you have to do."

"That worked. Thanks."

> "Well, from the poll results it looks like everyone has some experience with web conferencing. That's great."

As Cynthia continues talking, Gerri can hear voices in the background. Mary must have forgotten to mute her phone.

> Cynthia responds quickly. "Please remember to press Star 6 to mute your phone line. There are a few voices coming through." She continues, "I'm going to put up a slide with some terms that describe the interaction tools. Are there any that are new to you? If so, I'd like you to send a chat message. Use the chat box below my video."

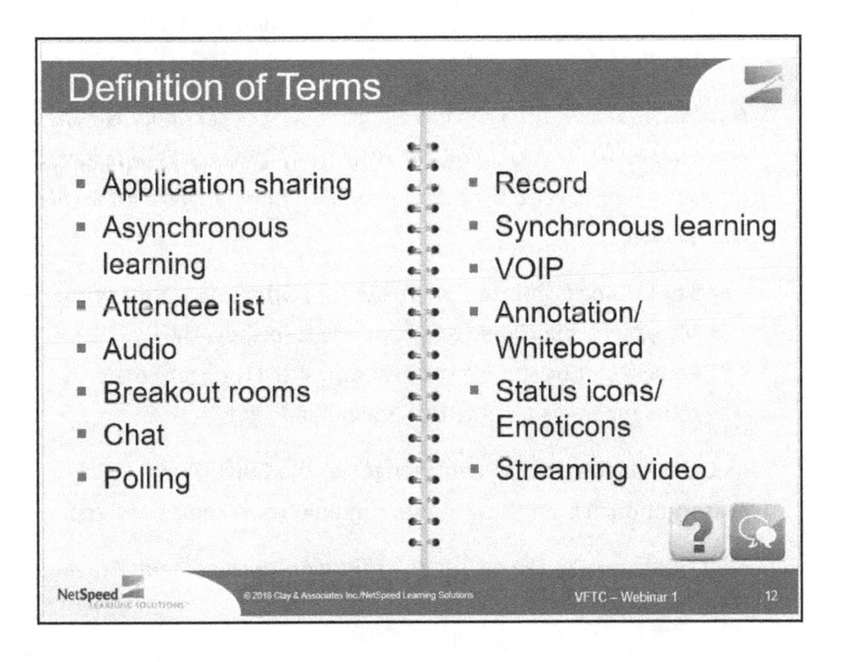

In the chat box, a participant asks what the whiteboard is for. Then comes a chat message from Mary: "My phone line just disconnected. I can't hear anything."

My goodness. Mary sure is having a lot of technical difficulties. I wonder if Tim will respond to her.

Sure enough, the chat box shows a message from Tim that says, "Dial back into the session using the phone number and pass code, displayed beneath this chat pod."

This is cool. Tag team troubleshooting, behind the scenes.

> Cynthia is continuing, "Okay, let's go on to the next slide. It goes into more detail about when to use VoIP and when not to."

The slide hasn't changed. Cynthia must have made a mistake. Now she has a strange look on her face. What's going on?

> "We seem to be having a technical problem that's freezing the slide presentation."

Not this again. They'll probably have to end the session. Just when I was beginning to have high hopes that this webinar is different. If it happens like this to Cynthia, I can just imagine what it'll be like when I give....

> Her musing is interrupted by Cynthia, who says, "No worries. My backup plan when this happens is to show the PowerPoint by sharing my desktop. We'll return to the web room when we want to use polling and chat."

As Gerri watches, the screen changes to the PowerPoint presentation in slide show mode, running from Cindy's desktop.

That was smooth, she thinks. Cynthia didn't miss a beat. Maybe I can do this after all.

Managing risk

This example included a variety of technical problems. Some were experienced by single individuals only, some by everyone. Some were trivial; one was a potential show-stopper. In all cases, the facilitator or host was able to easily correct them as they arose.

NetSpeed Note

The virtual classroom experience is significantly enhanced if you share the responsibilities with a host (aka producer). Rather than attending to technical details while facilitating a collaborative learning experience, consider having a host handle details behind the scenes. I'll emphasize this more in the next chapter.

Technology problems *are* going to happen to you. The best way to prepare for them is to follow three steps that are routinely used to manage risk. They are:

1. Anticipate the things that can go wrong.
2. Work to prevent them from going wrong.
3. When all else fails, make sure you know how to handle them if they do go wrong.

Let's walk through each of these steps.

Anticipating things that can go wrong

The first, and perhaps most important, advice I can give you for going into battle against technology snafus is to become comfortable with the features of your web platform. Understand what it can do and what it cannot. Take advantage of tutorials. Practice everything you will be doing: set up a chat window, clear it. Create a polling question, open it, remove it from the screen.

Research the differences between the host, presenter, and participant role. What can each of those roles do, and what are they limited from doing? If you need to give an instruction, are the menu hierarchies the same for you and the participant?

Web platform software allows you to change people's permissions. For example, you will often have to change an individual participant's permission before they can share content from their desktop. Make sure you have a level of comfort with creation of permissions.

Know how your Mute All function works. In some platforms, it mutes the presenter as well. If you, as facilitator, don't dial in using the facilitator password, the host may not know which attendee you are, and so won't be able to click Mute All without muting you as the presenter too.

After you've done your research and feel comfortable that you know your web conferencing platform's capabilities, brainstorm all the things that could possibly go wrong with them. Be creative. You might even try to break the tools to see what happens.

Enter your list of things that can go wrong into a Risk Management Plan spreadsheet, similar to the template shown in Appendix H.

Preventing things from going wrong

After anticipating what could possibly go wrong, the next step is to try to prevent those anticipated errors from actually occurring.

Let's return to a real-life mishap that I mentioned in Chapter 1. A presenter's training session dramatically ended because he mistakenly closed the connection. The web conferencing software would not allow anyone back in. While it's hard to make sure ahead of time that you won't sign off your session by accident, how might you ensure that the session wouldn't end? One idea would be to always have two hosts/presenters sign in on different machines. If one host is kicked off of one computer, the second host is able to keep the session open. (By the way, this solution might not work the same way with your company's web conferencing tool. Be sure to test it ahead of time.)

Another example from the first chapter related to hitting the maximum number of attendees that your web platform software allows. If you're running a marketing webinar and expect to hit the maximum possible, make absolutely sure that both your host and yourself have logged in before you open up the lines to other attendees. Come up with a foolproof way to keep from missing that important step.

NetSpeed Note

We mentioned that unlike web training workshops, marketing webinars can have hundreds of participants. As we have seen, this causes unique issues. Besides hitting the maximum number of attendees allowed by your platform, you may run into bandwidth problems. Read Appendix G for some important pointers.

Handling things that *do* go wrong

Unfortunately, the majority of technology issues that occur in web workshops simply can't be avoided. Enter step three of risk management: mitigate (that is, quickly correct or reduce) things that do go wrong.

Remember the potential fiasco in our example workshop? The slides that Cynthia was using within her web conferencing software suddenly froze. She would have been dead in the water if she hadn't anticipated this possibility and was ready for a backup plan: running PowerPoint from her desktop using Application Sharing. A backup plan is an example of a mitigation strategy.

It turns out that all of the other technology glitches in the Cynthia and Gerri example were similar in that there was nothing the facilitator could do to keep them from happening. Your participants will forget to mute their phone lines. Someone's phone line may disconnect. In essence, you have no control over things like this happening. The best you can do is anticipate them, immediately connect a symptom to a cause, and give instructions for resolving the issues. All without missing a beat.

Here are some other examples of things you don't have control over.

- *What participants do after you've turned over control of the screen*. Your learners enjoy playing with annotation tools, slides, clicking around the screen, exploring. This can wreak havoc on the learning experience. What's your strategy for regaining control?
- *A participant who doesn't have speakers on their computer.* This is a problem if you're using VoIP instead of a teleconference bridge. Without speakers, your participant won't hear the "voice" in VoIP. How should you have informed them of this ahead of time?
- *Bandwidth issues.* We mentioned ways to mitigate this. But you don't have control over your participants' desktop applications, their servers, or the inventory control module on their network that's taking up bandwidth. You just have to be quick at recognizing the issue is theirs alone, and then communicating that.
- *Some platforms require software installation.* With Adobe Connect, users don't need to load software ahead of time. That's not the case with other platforms. If your company's platform requires that attendees download software and they have issues with that, there's little you can do other than

making sure you've sent out multiple contact numbers for help.

Things you don't have control over include most of the scary things from the first chapter. Read it over again and pull these situations into your Risk Management Plan, then come up with a mitigation strategy. That way when you hear the forlorn call of a hapless student (e.g. my software doesn't work), you may be able to recognize the cause and identify that it's one of those things that, unfortunately, there's nothing you can do about it (e.g. "It could be your spam filter, popup blocker, or firewall. I'm sorry you're having difficulty. Try calling this number for technical support").

The breakout room blues

I've left the perils of the breakout room for last. This feature allows small groups of learners to collaborate amongst themselves, visited by the facilitator as needed, and then to return to the larger group when done with their project.

My message to you is: sometimes there will be technology breakdowns using breakout rooms. They will come from a number of directions. One has to do with how a participant originally logs into the session. If they call the teleconference (audio) bridge before logging into the conference on the web, the software doesn't know that caller is the same as the web conference viewer. When the breakout occurs, the individual will happily be part of it, but the audio won't go with them. In many platforms, instead of calling into the teleconference bridge, participants should log into the web conference and have the system call them.

Here are more:

- Many of your participants will work in companies where the system is calling a corporate switchboard. This means that the web system can't call them directly, which means they can't go into a breakout room. If you have participants from different companies in your workshop, you'll lose some and not others.
- Another issue: you send your learners to the breakout room, then pop in virtually to check up on them. If you have bandwidth issues, you can freeze the room.
- In some platforms, to get the breakout group to return to the mother ship with their whiteboard intact, one of them must have been given permission to save their work to their desktop. It's easy to forget to do that.
- In many platforms, someone other than the facilitator (usually the host) needs to set up the breakout rooms in the background; otherwise a giant lull in the action occurs while the facilitator adds each participant to the correct breakout room and everyone lapses into multitasking as they wait to hear what will happen next.

Breakout rooms are a great idea, but require practice to execute properly. My advice is to not use them if the class is meeting only once. But if you have an intact team that will be together over several sessions, it may be worthwhile to take them through the training that's required to ensure that everyone knows how to log in correctly, use the whiteboard and annotation tools, save a document to their desktop, and arrive back in the main room with their breakout room whiteboard.

To summarize this discussion of technology difficulties: do whatever is in your power to "expect the unexpected" so that you can either avoid problems or mitigate them when they occur in the classroom. This means learn the software, practice, be creative in your brainstorming, and be diligent in following good risk management steps. Good luck!

Homework

Assignment 8-1: Create a Risk Management Plan.

Read Appendix H and put together a Risk Management Plan following the steps described in this chapter.

chapter nine

getting it all together

Now, the moment you've been waiting for has finally arrived. You've written your learning objectives, worked out interactive and collaborative exercises, designed your PowerPoint slides. You've learned the intricacies of your web conferencing platform including all administration features, and feel comfortable that you can address whatever technology glitches the gods throw your way. Perhaps you've even taken a virtual facilitation course like the one offered by NetSpeed Learning Solutions (sorry, I couldn't resist a final plug). You're almost ready to apply your efforts to your first web workshop as a facilitator.

Just about. There are just a few final suggestions I'd like to offer, to ensure that your virtual facilitation career begins with the maximum chance for success.

Presenting an existing workshop for the first time

Your first step is to prepare the "meeting room." It's a check to make sure that your slides and assets like chat, polling, and whiteboard are complete and accurate.

This step may seem like overkill, but it's not. Have you ever **designed a poster, then sent your order for a thousand copies to the** printer? I have. More often than not, the content that you've painstakingly proofread contains a misspelled word (and it's usually a presenter's name). The instant the thousand fliers are published, the error just jumps out at you.

The concept is the same. Set up your assets on your conferencing software. Then walk through your presentation with fresh eyes for obvious errors like design, spelling, and missing slides; as well as for less obvious errors like inappropriate graphics or incorrect sequencing.

After you prepare the meeting room, your second step is to perform a dry run of the presentation with your host. Make sure that the two of you are crystal clear on your roles and hand-offs. Run through each slide and discuss the placement of polls and chats, confirm places where the host will take part in the exercise, and anticipate questions and concerns. A smooth dry run will give you added confidence and reduce anxiety. If things don't run so smoothly, or if during the run you decide to revise something, consider repeating the practice.

Unless you decide to do a dress rehearsal, this is your final check. Make sure you ask the host to keep her eyes open for the obvious errors that you're sure you captured during your meeting room preparation.

It's smart to consider a dress rehearsal if the web workshop is a completely new offering. Consider scheduling a pilot with five to six participants to work through all the exercises.

Even if you opt out of a dress rehearsal for your entire presentation, I'd recommend employing it for exercises that use breakout rooms. As we discussed in the previous chapter, breakout rooms can be tricky and unpredictable because you have to execute multiple steps correctly

Taking your time

Let's go back to the dry run. When we launch a brand new NetSpeed Learning web workshop, we allow a few hours for the dry run. If the slides or exercises don't quite work, you may have to redesign them. Or the host may make suggestions on how to tighten up the flow. The best practice is to schedule the dry run a few days before the actual event, to give you enough time to make changes.

If you've presented the same content multiple times, you should still launch a dry run because you never know when you'll run into unforeseen hurdles. It should only take ten minutes, and is well worth the effort. One time, my meeting event got corrupted; it had to be rebuilt and re-sent to participants. By stepping through a dry run the day before the event, I caught the problem before it had any negative consequence.

Employing a host

Since I haven't stressed the role of the host much in this book, let me take this opportunity to describe how your virtual classroom experience can be enhanced by sharing duties with a host.

The host (sometimes known as the *producer*) can handle technical glitches and logistical problems behind the scenes, allowing the facilitator to concentrate on her interactive presentation. Besides this key function, it's also fun to engage in a tag team relationship with your host: banter with him, ask him to share examples or ideas while people are typing chat messages, offer them the role of scribe in whiteboard exercises in which it's important for you to maintain control.

Consider the different ways the host and facilitator can divide the labor:

Role of Host/Producer	Role of Facilitator/Presenter
Send out web invitations	Send introductory email with pre-class assignment
Open training session	Take the lead in conducting the session
Review housekeeping and learning objectives	Engage participants in the use of the web interaction tools
Offer commentary, ideas, and examples in support of Facilitator's process	Share relevant stories and examples to meet objectives
Set up breakout rooms, if appropriate	Advance slides
Clear chat windows	Set up chat windows
Clear polling pods	Set up and run polling pods
Manage learner questions to Facilitator; answer logistics questions "behind the scenes" via chat	Answer learner questions out loud during session
Whiteboard comments	Whiteboard comments
Set up Application Sharing	Set up Application Sharing
Review and comment on chat messages	Review and comment on chat messages
Participate in role play or practice	Participate in role play or practice

Running solo

Am I suggesting that you shouldn't run a workshop session without a host? Yes. I believe the host to be that important, even if the only thing they do is handle technology issues. If you have to pause your presentation to address technical trouble, it could easily mean the difference between engaging your learners and losing them. You've been on the other side of the screen in web conferencing. One minute of time spent troubleshooting feels like ten minutes to participants (and an hour to you).

When you get more experienced, it's possible to do without the logistical duties of a host, such as clearing chat windows and polling pods. But even experienced presenters usually prefer the interaction that a host provides. One talking head presenter is boring. Banter between host and presenter perks up participant ears; engaging your host in some of the activities is a great way to energize the session.

Hosts are so important that I'd also recommend having someone else ready to step into the role in case of emergencies. Just as organizations often cross train facilitators, you should have backups for the role of host.

Scripting your presentation

PowerPoint contains a Notes section that is helpful as a guide. This is the place where you can paraphrase the content, include statistics or points that are not on the slide itself, and record your stage directions – that is, reminders on when to add and remove assets like chat and polling. Notes provide the additional assurance that you and your host are on the same page.

One best practice followed by NetSpeed Learning is to use a standard format that can be applied to all workshops by all presenters.

A standard format becomes familiar, and allows you to quickly process a large amount of visual information.

Here's an example of a Note that accompanies a slide from one of NetSpeed Learning's courses.

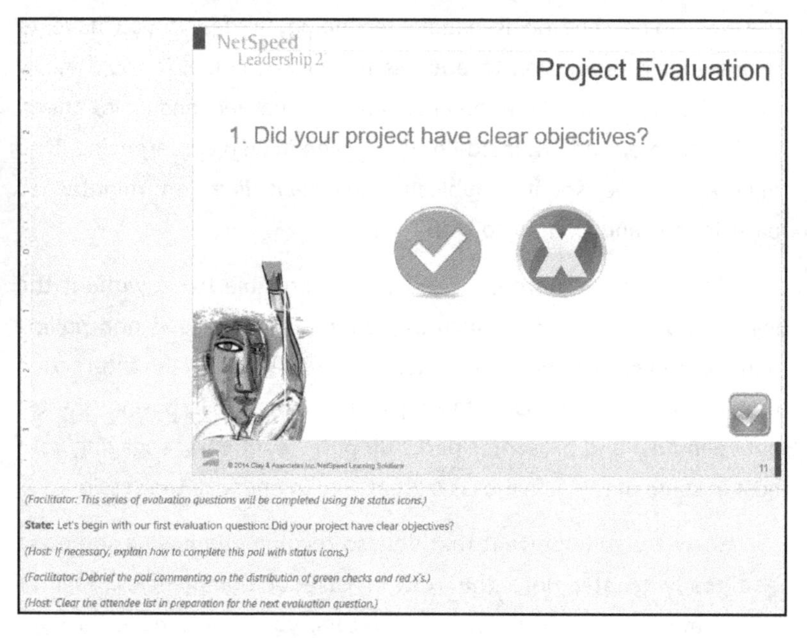

In this note, there are four sections. They each follow a convention:

1. A prep note for the facilitator, placed in italics and in parentheses.
2. A script for the facilitator, beginning with the word "State" in bold.
3. An action ("explain how to complete this poll with status icons") to be performed by host, if there is one, in italics.
4. An action for the facilitator, in italics.

We like to standardize on short phrases. You don't want to type out your script word for word or you'll end up reading it. Or, what's worse, you may start reading it to yourself.

Using a video camera

Throughout the book, I've emphasized the benefit of using a video camera. It doesn't need to be expensive, so there's no reason not to use one. We use a Logitech camera, but you can find an assortment of brands at any office supply store.

When you first purchase your camera, position it on top of your monitor, as close to eye level as possible. Then practice looking into the lens when speaking. Imagine you are making eye contact with people. Get comfortable with it. Make sure the background looks presentable and professional. If your camera has a zoom feature, play with different settings so you'll know what range is most attractive. I recommend a setting where your head and shoulders are in the camera window and there is limited background detail.

After your initial workshop, continue to test your camera ahead of time every time you use it. Do this even if you've never had problems in the past. The same goes for testing your VoIP and headset.

Sometimes a working camera can act to your disadvantage. Once I presented a workshop after lunch. To my horror I noticed there was something in my teeth that greatly resembled the spinach I had eaten at lunch. When I reached a point where my participants had taken over conversation, I discreetly muted my mike, froze the camera, and removed the offending vegetable.

Lesson learned: before each event, get into the practice of using your video camera to check your general appearance, or have a mirror handy.

NetSpeed Note
Technical tip: Even inexpensive webcams can produce satisfactory images as long as the subject has plenty of light. In most office configurations, we're facing a wall or corner. The only light that shines on our face is the dim blue glow from our computer monitor, while the background behind us may be brightly lit. To avoid looking like a silhouette, I use an inexpensive fluorescent desk lamp positioned just near my webcam. Make sure the light is a few inches above eye-level, to avoid those Halloween shadows.

Web platforms are not created equally

There are differences among the web platforms that will affect how you step through your meeting room preparation, dry run, or dress rehearsal. Become familiar with how you schedule a meeting or event in your platform. For example, although Adobe Connect has no limits on how early you can get into your meeting room, some platforms require scheduling the presentation thirty minutes ahead of the actual start time or you can't get in.

For your prep or dry run, some platforms allow you to practice on the event link that you created for the workshop; others consider the link to be a one-time-only affair.

Platforms also differ on when you can place assets in the meeting room. Blessedly, Adobe Connect allows you to create all assets in the room ahead of time. Others allow you only to prepare offline and then add the polling question or whiteboard to the meeting room right before your event. For these platforms, preparation is even more important.

I *really* prefer web platforms that allow you to set up everything in advance. If you have any input into your company's platform tool, make sure this feature is considered highly important during the evaluations. It may save you from embarrassment when you're "live."

Speaking of embarrassment, I'm reminded of another lifetime when I was acting in a one-woman play. The scene featured a key prop, a compact mirror. The raising of the curtain found me buried to the waist in a mound of sand. When it came time to use the mirror, I realized it was nowhere to be found onstage. Mired in sand, I mimed looking at my reflection in the palm of my hand until my stage manager finally ambled onto the stage and rescued me. The moral of the story – "check your props" – is just as appropriate in the world of virtual facilitation.

Creating the training event

Most web platforms have tools for creating the event, sending invitation emails, accepting registrations, and sending reminders. The invitation should include login information, dial-in information and passcode, a point person for questions about the event, and, perhaps a phone number for technical questions in case something goes wrong. Clarify with your team who has responsibility for this email – whether it is you, your host, or someone else. You don't want this falling through the cracks.

Make sure someone on your team is watching for emails on the day of the event. It's not unusual for our host to receive five emails on the morning of the event from participants who can't find the email with instructions, have trouble logging in, or have other concerns.

Assuming your host is going to tackle technical problems in the background during the workshop, he should have the tech support phone number handy if needed. I like to print out the pages of the user manual for the most common things that I will be doing, with useful text highlighted. Yes, I may know how to mute or unmute the phones, for example, but in the heat of the moment it's easy to forget.

Another handy item is the risk management sheet that you developed (from the previous chapter). Use it for reference when emergencies arise.

On the day of the actual event, you don't want to wait until fifteen minutes before the start time to discover there's a problem. Instead, log in a full thirty minutes early. Test everything we mentioned in the sections above. If it all checks out okay, great. You can always put yourself on hold for twenty minutes and go get yourself a latte. (Just check that mirror again for a whipped cream moustache.)

NetSpeed Note

This chapter has covered the steps you need to get your web workshop finely tuned before formally presenting it to an authentic audience. We've provided a logistics checklist in Appendix I as an easy reminder. There are also some sample questions and steps in Appendix J that cover everything you've learned about how to design an interactive, collaborative learning experience.

Parting words

You've done it! We're at the end of our suggestions, war stories, best practices, and illuminating insights. You have all the tools you need to begin to deliver interactive training that captivates, informs, and energizes. As you embark on your web workshop training, please use this book as a reference and visit us at www.netspeedlearning.com to connect with other readers and share your tips and tricks. Now it's time for you to go out and deliver Great Webinars!

Homework

Assignment 9-1: Design and present your web workshop.

Use the steps in Appendix J to complete your web workshop design. Practice your presentation, then perform it. Good luck! Call us if you need us – we love to coach new web facilitators.

chapter ten

Brain-based Principles for Engagement and Retention

"The feeling of connection stabilizes and propels a person. It promotes growth. Without the invigoration of connection, the brain shrivels and life sags."

"In the absence of connection, fear usually rules."

– Edward Hallowell[xxiii]

The two thoughts above inspire me as I think about the strengths and weaknesses of virtual training. In our fast-paced, technology-fueled workplace, it can be all too easy to overlook the power of connection with everyone on our teams. Researchers have confirmed in studies over the years that social isolation and the lack of personal connections result in poor health and early death. As organizations move toward flexible work schedules, home-based offices, virtual teams, and virtual meetings, it is essential that we attend to people's need for connection.

I'm a continuous learner. If you've read this far, I'm betting you are too. As the technology that supports virtual classrooms and asynchronous learning elements has improved, I have focused on designing and delivering even better digital experiences. With the emergence of microlearning and social learning, our team at NetSpeed Learning Solutions has evolved an approach to talent development that capitalizes on the preferences of today's learners.

When we capitalize on the way our learners' brains take in and process information, we significantly improve our virtual learning programs. The cognitive research has well documented that we need contrast and comparison, stories and metaphors, moderate stress, emotional connection, visually compelling graphics, social collaboration, and the opportunity to bring our personal context to the content if we want to learn anything.

Neuroscientists have been exploring the amazing neuroplasticity of our brains. Simply put, our brains are constantly changing and adapting. We want our participants' brains to light up like neon signs as they build new neurons and synaptic connections. The very act of learning, practicing, and applying new knowledge changes the wiring of our brains.

Don't believe it? From the research being done in positive psychology[xxiv], we can observe the result of intentionally engaging and rewiring people's brains. Someone who has developed the habitual pattern of pessimism might begin a gratitude practice. Consciously writing down five things that happened during the day for which they are thankful gradually develops and reinforces new neural pathways. Within 30 days, the habitual pattern of pessimism may be replaced by a pattern of optimism fueled by gratitude. I find that awe-inspiring.

As virtual learning professionals, let's adopt the attitude that we are helping our learners create new neural pathways. It is both a privilege and a responsibility to guide that process. For me, it all begins with engagement in the virtual classroom and the first six brain-based learning principles.

Brain-based Principles for Maximum Engagement

- Active Engagement = Active Brains
- Neurons that Fire Together Wire Together
- Vision Trumps All Other Senses
- Social Learning Fires Mirror Neurons
- No Pain – No Gain
- Practice Makes Permanent

Active Engagement = Active Brains

In my virtual training sessions, I often ask people to rate their ability to actively engage people in the virtual classroom. This brain-based principle reinforces the need to engage your learners every two to three minutes. Any time you find yourself developing a lecture-based presentation, stop and ask yourself, "How can I actively engage my learners' brains as they are exposed to this content?" Use deeper questioning techniques that encourage evaluation, analysis, problem-solving, and reflection. Successful designers and facilitators recognize that lecturing for ten minutes just doesn't work well to actively engage people's brains.

Here's an example from NetSpeed Learning's Virtual Leader™ program: We facilitate a set of activities to help leaders determine how they might need to modify their leadership style to work more effectively in the virtual workplace. We introduce Theory X and Theory Y, based on the work of Douglas McGregor, a professor at Sloan School of Management at MIT.

Of course, we first need to explain what that theory is about, but we do that in a short "lecturette" that tees up a chat discussion. Briefly, Theory X managers believe that employees are inherently lazy, are not happy with their jobs, are more motivated by punishment than rewards, and need to be supervised closely or they won't perform. In contrast, Theory Y managers believe that employees are inherently self-motivated, ambitious, want to perform well at work, and are able to exercise self-control. It's helpful to think of these descriptions as holding down two ends of a continuum with Theory X at one end and Theory Y at the other.

After that quick introduction of Theory X and Theory Y, we ask participants to consider how a manager who holds Theory X beliefs would manage their virtual team and how a manager who holds Theory Y beliefs might manage their virtual workers. We open up two chat pods in Adobe Connect, and have them think about the implications of holding these beliefs on the leaders' actions.

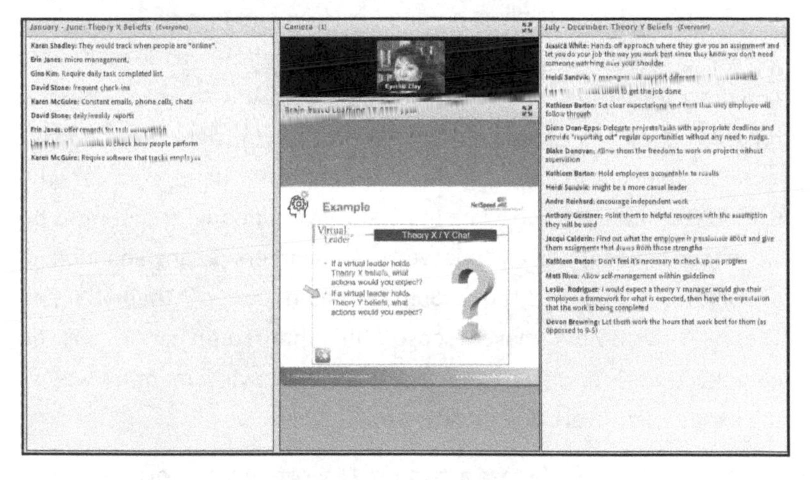

Frequently, participants chatting about these behaviors reveal that they much prefer the Theory Y leader's approach to management because they believe it is more empowering and respectful. We then ask them to complete a poll where they identify their own leadership approach. Are they more reliant on Theory X or Theory Y leader behaviors? It's not surprising for me to see that most people say they rely on "75% Theory Y and 25% Theory X" behaviors, as if that might be the correct answer (as you can see in the screenshot below).

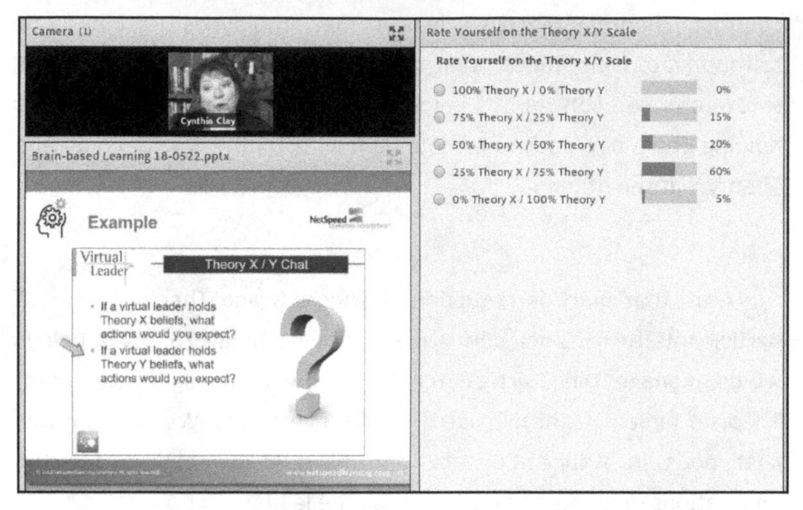

Surprise! There really is no one best combination of Theory X or Theory Y behaviors. In the virtual workplace, a leader who demonstrates 100% Theory X behaviors is likely to micro-manage (because they can't see what people are up to), demotivate virtual workers, and create disengagement. On the other hand, a leader who uses 100% Theory Y behaviors (basically giving people freedom to do what they choose) is likely to leave virtual workers feeling abandoned, disconnected and frustrated. Some combination of Theory X and Theory Y behaviors makes sense, but that combination will be influenced by your organization's virtual work environment, as well as the experience levels of your virtual employees.

Notice the flow of these activities: 1) short introductory lecture, 2) questions to encourage reflection and sharing of opinions, 3) self-assessment, and, finally, 4) the key learning point. Of course, we could have just lectured everyone about what beliefs and behaviors would be demonstrated by Theory X and Theory Y leaders and spoon-fed them the key learning point. It might be quicker, but our learners would likely be multitasking rather than engaging. In other words, there would be no neurons firing in those brains!

Neurons that Fire Together Wire Together

Neuroscientists have long noted that "Neurons that fire together wire together." The more associations we can make linking our topic with stories, metaphors, thoughts, feelings, or kinesthetic sensations, the stronger the neural network that we create.

Think about the most memorable learning experience you've attended recently. Chances are good that when you recall that experience, it comes with the memory of a speaker or a colleague in the class sharing a compelling story. As you retrieve those stories from your memory, you also access the core concepts of the training.

To illustrate this brain-based principle, I often share the experience of an elementary school teacher who was trying to help her young students learn to read aloud. Sitting at their school desks, they read the same passage out loud together as she tried to explain punctuation. Now, if you have ever listened to young ones read aloud, then you know that they tend to skip over commas, periods, and exclamation marks, reading the words like one long run-on sentence. Despite her coaching, after 30 minutes of practice, these kids had made very little progress.

Then this inspired teacher had them put on their coats and go outside to the playground. There she arranged them in a circle and explained that they would walk clockwise as they read aloud together. When they came to a comma in the middle of a sentence, they would pause and slowly take their next step. When they came to a period at the end of the sentence, they would stop walking (full stop). And when they arrived at an exclamation point at the end of a sentence, they would jump up and down with excitement. After 15 minutes of active practice, she brought them all back inside and was delighted to see the level of improvement in their ability to read aloud with dynamic punctuation.

How does this story demonstrate the brain-based principle, **Neurons that Fire Together Wire Together**? What my students often notice is that learning occurred through multiple modalities. They were reading the words on the page (visual), speaking the sentences out loud (auditory), and engaging their bodies (kinesthetic). They were learning in a structured, playful way in a comfortable environment (the playground).

To leverage this brain-based principle, I encourage trainers and facilitators to share personal experiences that support the key learning points. Also write juicy questions that elicit success stories and challenges from participants. Go beyond rote memorization and repetition of information by adding storytelling to the mix. Activate multiple sensory modalities and get those neurons firing and wiring.

Vision Trumps All Other Senses

I want to acknowledge that this brain-based principle is one of John Medina's brain rules[xxv]. Learning professionals used to think that individuals had different learning preferences based on their dominant sense (visual, auditory, or kinesthetic). I was once firmly convinced that I was a visual/kinesthetic learner. But neuroscientists have since debunked that old notion. For most of us, vision is the dominant sense; yet information that is presented and reinforced through multiple sensory channels is more likely to be retained.

If you understood the importance of vision as the dominant sense, then you wouldn't make the mistake of simply opening up the teleconference bridge and having people talk about topics. You would leverage your web conference platform as a rich, visual medium. I love Adobe Connect because this web conference platform allows us to create multiple layouts and place chat pods, polls, whiteboards, note pods, and share pods in different configurations. We can resize and layer these pods on top of each other. Every few minutes, we can visually change the look of our participants' screens to engage their attention and interest. Other platforms seem static and boring in comparison.

Whatever web conference platform you are using, however, deliver your content with compelling graphics and images. We discussed the best use of PowerPoint in chapter five. Think also about how you can incorporate microlearning elements in the form of video before, during, or after the virtual classroom experience. Rich graphics, animation, and appropriate use of text on the screen contribute to memorable learning nuggets.

You might want to explore software that allows you to create infographics. My personal favorite is Piktochart. With pre-designed templates and icons, this web-based application allows people with little graphic design experience to create pleasing visual reinforcement tools. Creating a decent infographic challenges you to distill your thoughts to central messages with graphics that lead the learner from Point A to Point Z. Often it helps to include a "call to action" and guide your participant to take a next step.

NetSpeed Note
Once you've completed this chapter, download an infographic. We created this visual tool as a reinforcement of the first six brain-based principles for engagement. You'll find it at our website: https://netspeedlearning.com/contact/?p=nsi&r=brain-based-learning

Social Learning Fires Mirror Neurons

Scientists used to think we acquired the ability to recognize and respond to emotions slowly as our brains matured from infancy to childhood. What they now recognize is that babies begin to scan the faces of their caregivers within a few hours of their births, taking in emotional cues almost immediately. Mothers and fathers everywhere probably recognized this phenomenon well before neuroscientists proved it. I remember holding my newborn daughter when she was two days old, feeling her penetrating gaze on my face. When you watch someone experience an emotion, your brain fires mirror neurons in the same pattern[xxvi]. If they feel happy, your brain mirrors happiness. If they feel disgust, your brain mirrors disgust.

One implication of the brain-based principle, **Social Learning Fires Mirror Neurons**, is that our learners' brains will fire mirror neurons that reflect the learning environment we create. Showing up on web camera with positive, enthusiastic virtual presence can help you create a lively, engaging virtual learning session. On the other hand, we need to do everything we can to avoid creating negative emotional states for our participants. That's just one reason why reacting in anger or frustration to a technical failure during a web workshop should be avoided at all costs.

The 19th century British playwright, Arnold Bennett, once observed, "There can be no knowledge without emotion... To the cognition of the brain must be added the experience of the soul." In other words, learning occurs when we trigger positive emotional states and memories, connected to new knowledge. Learning is essentially a social, collaborative, human experience.

We learn from each other, sometimes without even consciously realizing that we are doing so. Scientists have studied the brains of people while they observed someone performing a task or learning a new skill. Without actually performing the task themselves, their brains fired mirror neurons as if they were actually doing the work. Sports psychologists know that the tennis player who visualizes swinging their racquet correctly will see similar improvement as the player who actually practices swinging their racquet. Does this awareness mean that a tennis player never has to play tennis to master hitting the ball consistently? Of course not. But it does encourage us to design virtual learning experiences that include observation and visualization as well as practice.

Let me tell you a story. Several years ago, we raised a small flock of newly-hatched ducklings in our backyard. They used a shallow pan of water in which to bathe and drink. As they quickly grew into big ducks, they outgrew their pan. Feeling sorry for them one day, I filled up a small barrel of water and put a ramp on the side. The first brave duck, who my daughter named Hope, walked up the ramp and plopped into the barrel. Until that moment, she had never done anything except stand in a shallow pan of water. Now she attempted to stand in that barrel which was a good two feet deep.

For several minutes, Hope appeared to be drowning as she dropped down in the barrel, trying to touch the bottom, and then resurfaced with frantic quacking. The other two ducks circled the barrel while this drama was occurring with consternation on their faces. I confess that I was worried too. What if I had just drowned my duck? Finally in exhaustion, Hope surrendered to her fate and stopped struggling. What do you know, she floated! She quacked happily in surprise and paddled around the barrel. I could swear she was smiling with satisfaction. That was all it took for the other ducks to take their turn waddling up the ramp, plopping in the barrel, and paddling happily in the water, imitating their brave sister.

Like our flock of ducks, people learn from observation, visualization, and practice. I'm sure those mirror neurons were firing in their feathered brains, just as they fire in our learners' brains.

I told this story once to a group of resistant trainers who were learning how to deliver engaging virtual learning sessions through our Virtual Facilitator Trainer Certification course. These folks did not want to deliver their 20-minute trainback webinars. Their arms were crossed as they protested the assignment. They assured me that they didn't need or want to practice. The truth is they were afraid to fail in front of their colleagues. I reminded them that they had tons of experience as classroom trainers. Hope was a hardy duck who learned to float in a new environment. They were trainers who were learning to facilitate in a new environment. And like Hope, they were not going to drown – they were going to float!

Thankfully the first trainer delivered her trainback while her colleagues observed and learned from her example. We applauded her success and acknowledged her strengths. One by one, each of them proved that they were ready to transfer their previous experience into this online environment. Social learning was actively firing those mirror neurons!

To leverage this brain-based principle, consider how you will stimulate positive emotional engagement during your web workshop. Plan to keep yourself on web camera throughout the session so your enthusiasm and positive energy spark mirror neurons. And think about how to create peer-to-peer collaboration with observation and practice in the virtual classroom.

No Pain – No Gain

Like Hope the Duck, if you never leave your comfort zone, you're unlikely to learn anything new. The brain-based principle, **No Pain – No Gain**, describes the need to nudge people out of their familiar ruts, if they are going to master new skills. Essentially, failure precedes success. The trick is to create a learning environment and process that allow people to fail safely.

Let's face it. Most adults hate appearing to be incompetent or unskilled. It takes courage to step up and try something for the first time, willing to do it badly before doing it skillfully. We wish we could just watch a YouTube video and go perform the task with mastery. As virtual instructional designers and facilitators, we must create learning experiences that gradually require more of our learners. We might begin with simple chats or polls to warm people up. During the session, we might offer video demonstrations that people can watch and critique. By the end of the web workshop, we might ask people to role play or practice on web camera and receive feedback from their peers.

Remember the Goldilocks Rule: Too much failure shuts down learning. Too little failure bores people. Just the right amount of failure increases curiosity, challenge and learning. In the virtual classroom, our participants need to be stretched by relevant issues, challenges, and cases so that they develop new skills. I use this cute cartoon (a slightly modern take on the fairy tale) in one of my virtual training sessions (licensed from Deposit Photos).

You might begin a training session with a quiz to let people test their own knowledge of a topic. Or you could give them a case study problem and have them offer ideas and suggestions about how to resolve it upfront. Let them know that you will return to the same case problem at the end of the web workshop to see if they might apply what they learned to create a different solution. To leverage the **No**

Pain – No Gain principle, challenge yourself to challenge your participants and move them out of their familiar comfort zones.

Practice Makes Permanent

The sixth brain-based principle for engagement is **Practice Makes Permanent**. Notice that it's not **Practice Makes *Perfect***. The quest for perfection shuts down the learning process. As you design a blended virtual training experience, you must start by identifying your terminal learning objectives (what people will actually be able to do when they return to their work). The only way to achieve those objectives is to ensure that people actually practice during and after the virtual learning event.

The seven-step process can be described this way:

1. Review steps
2. Model or demonstrate
3. Have participants reflect
4. Apply (practice)
5. Observe and share feedback
6. Use on the job
7. Report results to peers

Notice that there is practice built into the web workshop experience, but the learning isn't complete until there is practice on the job. In our online courses, we build a feedback loop into the design that requires participants to apply it on the job and then report their results back to their peers in NetSpeed Fast Tracks, our social collaboration website.

Application and practice are vital aspects of effective virtual learning design. You might be tempted to skip them, but this is where the rubber really meets the road. For example, If you are teaching a group of remote leaders how to delegate during a virtual classroom session, they might not be able to actually delegate a task to someone during the synchronous virtual session, but they will be able to move through the seven steps in this way:

1. Have them download a delegating template with steps to follow for the virtual workplace. Review the steps.
2. Review and discuss a completed delegating worksheet. How is it different than delegating when you're face to face?
3. Have managers identify a task they want to delegate and decide to whom they will delegate it.
4. Give them time to complete the delegating worksheet so they prepare to delegate this task when they return to work.
5. Have one or two participants share their delegating worksheets and receive feedback.

6. After the web workshop, ask managers to delegate the task using the delegating worksheet.
7. Have people report what they learned from the delegating experience back to their colleagues.

Leverage the brain-based principle, **Practice Makes Permanent**, to ensure that your virtual training sessions actually result in skill development.

The first six brain-based principles for *engagement* are the foundation of great virtual learning. They work in partnership with six more brain-based principles for *retention*, described in the following sections.

Brain-based Principles for Maximum Retention

- Prime the Pump
- Chunk It Down
- Mix It Up
- Sleep on It
- Test and Retest
- Use It or Lose It

Prime the Pump

The brain-based principle, **Prime the Pump**, refers to an old-fashioned water faucet in which you have to pump the handle to move the air out of the system before the water can flow. I use this image to help us understand how we need to prepare our participants for a virtual learning experience.

I recently registered for a course that was being taught by a live instructor online. I received an email the day before the course with my log in instructions, the course title, the date/time of the course, the instructor's name, and directions to purchase the book required for the course and read the first chapter.

I did as instructed and completed about half of the reading before logging into the first session of a multi-week program. Frankly, I stopped reading because I was bored. As I pondered my lack of excitement and interest, I realized that a core brain-based learning principle was missing: **Prime the Pump**.

I had received detailed instructions for attending and participating but none of my motivational or social (affective) needs had been addressed. What would I gain from participating actively? Who else would I be learning with? Would we be working together? Why would this course be a valuable use of my time? Has it changed anyone else's life or work? If so, how? Why was the instructor the right person to be leading the course? Was she qualified? Would I like her? Why would I come back week after week?

When we **Prime the Pump** for our participants, we open the door to a vibrant learning experience. We lower the barriers that may prevent people from participating actively due to previous negative educational experiences. We help people reflect on their personal goals and connect the objectives of the course to their needs and interests. Indeed, technology makes it possible to bring learners together in social, collaborative groups to introduce themselves before the course even begins. Trainers and facilitators can welcome participants with warm, introductory videos a few days before they log into the live session.

When you **Prime the Pump**, you put yourself in your participants' shoes and consider their underlying needs, interests, and motivations. What are you doing *before* your participants arrive in your virtual classroom to engage their interest and attention? You want to ignite their desire to learn and prepare them for something new before they log in. Think about using a short video, a puzzling on-the-job challenge, or a self-assessment. Imagine telling the first part of a dramatic story in a recorded video, and then inviting them to come to the web workshop to get the rest of the story.

A handy memory tool to get participants ready is thinking about their ABCs. A stands for Attitude (Affective). B stands for Behavior. And C stands for Cognitive. (You were introduced to these ABCs in chapter three.)

In preparing people to learn, we need to consider their Affective needs. Write a warm email or other notification that sets the stage for a personal, connected learning experience. Introduce yourself and, if possible, start to make connections between participants.

As you prepare them, focus on the Behaviors or skills they will master. Communicate *terminal learning objectives* that describe what they will be able to do as a result of participating actively, not the activities that will occur in the virtual classroom. Consider including a short video with your first welcome email featuring a respected colleague describing what they learned and how it helped them tackle a problem or achieve a goal.

For Cognitive needs, prepare participants by having them check their knowledge. Can they solve a tough case study? Do they know how to resolve a difficult on-the-job challenge? Promise that by the time they leave the virtual workshop, they will know all they need to know to handle these tough issues.

Chunk It Down

Chunk It Down is a brain-based principle that reminds us that shorter is better, especially with the dwindling length of attention spans we are experiencing. The question is "How short?" When I ask participants in my web workshops and presentations, "What is the ideal length of a content segment?" I hear responses like, "as short as two minutes" and "as long as 20 minutes." The trend toward microlearning may lead you to think that all content segments must be as short as two minutes. Certainly, that is probably true for microlearning videos and podcasts. In those cases, it's helpful to create one objective and provide the content that fulfills that specific objective.

In the virtual classroom, you can also incorporate microlearning elements. For example, you might show a short video preceded by a thought-provoking question. Ask participants to watch the video and be ready to answer the question in chat. I believe that we can actually sustain people's attention in the virtual classroom, on one topic, for 15 to 20 minutes. The only caveat is that those 15 to 20 minutes must be interactive and compelling. Shift their attention and interaction every two to three minutes. If you have them engage with you, with the content, and with their peers actively, those 20 minutes will fly by. As you design a web workshop experience, you can probably include two or three topics in a 45-minute session or three to four topics in a 60-minute session.

Here's a sample process that we use in a 15-minute section of the Virtual Leader program to help managers think about the motivation of their virtual teams:

1. Show a three-minute video featuring an inspiring small business that is making a difference in their community.
2. Pose a discussion question to have people respond in chat (tee up this question before the video): *What is the most important takeaway as you think about your team's motivation?*
3. Debrief the chat responses, looking for themes and patterns.
4. Make the key learning point (and connect it to their chat comments): *For many people, work has little value unless it contributes to a higher purpose.*
5. Have managers reflect and respond in chat to the question: *How will you adapt your actions as a virtual leader to help people understand the purpose of their work?*

In this example, we focus on one core topic (the need for the virtual leader to create a strong connection to the team's purpose).

Notice that each individual activity in support of this topic takes two to four minutes. We move through a logical sequence: watching a video; discussing a question about motivation; making the key learning point; reflecting on their actions as virtual leaders. This is what I mean by **Chunk It Down**: short learning activities, in support of specific learning objectives, that take place in a 15 to 20 minute segment in the virtual classroom.

Sometimes a client will ask us to record a webinar so that people can view it as a substitute for attending the live, synchronous webinar. Watching people interact through chat and polling is not at all the same as participating live. However, if we do agree to that request, we always edit the webinar recording – removing silences, pauses, and those moments where the facilitator is waiting for people to answer a discussion question in chat. We tighten up each segment so chat responses appear instantly. We reduce the webinar recording segment to seven to ten minutes per topic. If a polling question is provided during the webinar recording, we add that polling question in NetSpeed Fast Tracks, a social learning platform, so that they can share their opinion and compare their responses to those provided by others who are watching the asynchronous recording segment.

In the quest for ever-shorter learning segments, we don't want to sacrifice interaction, if we expect people to remember and retain the content.

Mix It Up

You've probably heard about the forgetting curve, the idea that we forget 90% of what we've learned within a week of the classroom experience. Forgetting is an adaptive function performed by our hippocampus. Our hippocampus filters the plethora of stimuli coming at us to identify what is most important to pay attention to in our environment. Without its focus on prioritizing stimuli, we'd likely be swamped by information (and possibly immobilized).

Here's the trick: the hippocampus has been designed to filter out irrelevant, redundant information. Constantly scanning the environment, this region of your brain is always questioning whether it should move the next bit of information into short-term memory or ignore it. Because we can't hold much information in short-term memory (about seven bits[xxvii] at a time), what is there degrades over time. Something has to be bumped out to make room for something else to stick. This is why we tend to recall what happened at the beginning and at the end of a training session. Everything in the middle can seem like a blur.

If you've ever been sitting at your desk watching a slow-paced virtual classroom session, you might have suddenly realized that you were day dreaming. Thank your hippocampus, the region of the brain that has determined that you don't need to pay attention. The brain-based principle, **Mix It Up**, suggests that as facilitators, we need to bypass this adaptive function in creative ways, to get and keep people's attention.

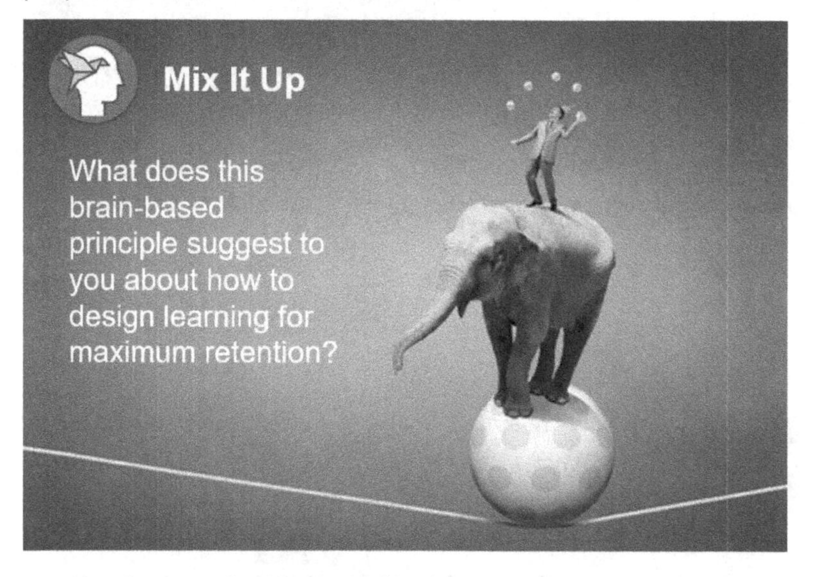

The keys to helping people attend to and retain important information are variety, novelty, mixing, and spacing. As you design and facilitate synchronous web training, use every interaction tool in your web conference platform. Break the linear approach to design:

Blah, blah, blah, blah. Chat.
Blah, blah, blah, blah. Poll.
Blah, blah, blah, blah. Chat

Instead, create activities that combine multiple interaction methods in innovative ways. Is it more challenging for you as a facilitator to lead a virtual session with lots of variety? Indeed it is! But your participants will thank you because they want to participate actively.

Recently I was speaking with a client who praised the performance of one of our Virtual Master Trainers, telling me that her average course evaluations for an interactive online course we had designed for them were the best of dozens of virtual courses (4.8 on a scale of 5.0) that they led in this Fortune 500 company. In the next breath, he complained that when he led the course himself, it took him a long time to prepare and he was exhausted after two hours of delivery. I pointed out that it was no accident that her ratings were so high. She was bringing variety and engagement to the experience. It's a practiced skill to lead a web training session that is based on the brain-based principle **Mix It Up**.

The next variable to consider is novelty. I fly a lot to different locations so I can deliver workshops and speak at conferences (ironic, I know, since I often speak about how to deliver engaging webinars). I sat down on the plane, buckle my seat belt, and ignore the flight instructions. I've heard them so many times my brain isn't the least bit interested. Every once in a blue moon, a savvy flight attendant will get my attention with witty comments. My ears perk up and I actually listen to every word, laughing along with the rest of my fellow passengers. One Southwest flight attendant (you can find her on YouTube) gives hilarious instructions, including the memorable line, "If you are flying with young children (pause), we're sorry. If the oxygen masks fall, pick which child has the greatest earning potential and put their mask on first."

A few years ago, I was flying from Seattle to Spokane. Twenty minutes after we took off, our pilot came on the loudspeaker, "Ladies and gentlemen, we have found a crack in our windshield. I'm sure you will agree with me that it is best to have a pilot onboard your plane." Everyone laughed. He continued, "We will be turning back to Seattle to get a new aircraft, one that is capable of keeping its pilot in the cockpit." He assured us we were in no danger as he turned our plane around and calmly flew back to Seattle. Novelty and humor were both at work in that interaction. He disarmed any fear we might have felt at the news that our windshield was cracked, with dry humor.

Novelty has that same effect in the virtual classroom. As you design your virtual sessions, think about how to surprise people with a creative twist or turn. Avoid a plodding, predictable flow, even if you think your topic might be plodding and predictable. Bring a creative approach to your activities as you use a variety of interaction tools to help make your learning points.

Mix It Up also includes paying attention to how you practice. In an article that appeared in *Training Magazine*[xxviii], the author, Randy Sabourin, described a study conducted at Cal Poly in which the baseball team was divided into three groups to practice hitting fast balls, curve balls, and change ups. They were interested in understanding which kind of practice yielded greater proficiency in hitting curve balls. The control group took no additional practice. The second group practiced hitting 15 fast balls, 15 curve balls, and 15 change ups in blocks (let's call it blocked practice). The third group practiced hitting a fast ball, a curve ball, and a change up, 15 times – never knowing which kind of pitch they were about to receive (let's call this interwoven practice). They conducted extra practices twice a week for six weeks for Groups 2 and 3. At the end of the six weeks, both groups showed improvement. The group that used blocked practice improved nearly 25%. But the group that used interwoven practice improved by nearly 57%.

I find it interesting that Group 2, the group that relied on block practice, found the sessions less frustrating because they could see the short-term improvement of their batting skills. Group 3 found the practice sessions much more challenging (and less satisfying), but in the end, achieved much higher long-term skill improvement.

The other variable at work in this example is spacing. By returning again and again to recall and practice a specific skill, both groups experienced significantly greater improvement than the control group. They experienced a piddling 6.2% improvement over the same time period because they had no extra spaced practices.

We know that it takes two hours of repeated practice to build the synaptic connections in our brains that support the development of a new skill. Of course, you are not likely to spend two hours in a single virtual classroom session practicing, but you can build practice into the design of a complete learning experience by ensuring that people take time to practice after the synchronous classroom event. You might schedule short practice sessions spaced out over several weeks. Then follow up to have participants report on their progress.

Sleep on It

Sleep! You probably agree that we all need enough sleep to function well. When we don't get enough Z time, it can impair our productivity and accuracy. But did you also know that sleep is essential for our brains to build long-term memories[xxix]?

The student who crams all night for an exam the next day might be surprised to learn that catching even a few hours of sleep right after studying will do more for increasing knowledge retention than staying up all night without sleep. In those hours of sleep, information stored in short-term memory moves from the hippocampus to the neocortex. Scientists used to think that REM (rapid-eye movement) sleep was where all the action happened. But they now understand that SWS (slow-wave sleep) is critically important for memory consolidation. The good news for people who love to learn? We spend more time in SWS then we do in REM sleep at night.

As training professionals, we need to understand how we form memories and how we forget memories. The four phases of this process might be summarized as:

- Learn it
- Encode it
- Recall it
- Apply it

Simply stated, the waking brain (hippocampus) encodes memories in short-term memory. The sleeping brain (neocortex) integrates memories or consolidates them into long-term memory. The recall or retrieval process strengthens long-term memory and increases the speed at which we can access that information. And we need to apply it after we recall it. That's why "practice, practice, practice" is so important for anyone learning a new skill.

If we take a one-hour nap after learning something, or go to bed for the night within an hour or two of reviewing critical facts or information, our brains get to work, consolidating that information into long-term memory. We dramatically increase the chances that we will be able to retrieve those memories the next day. And the act of retrieval then strengthens long-term memory.

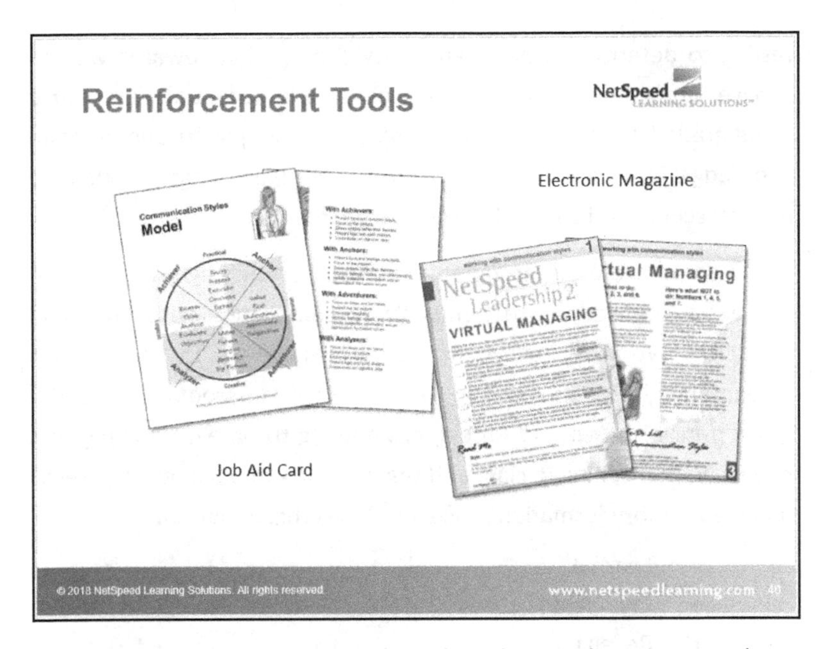

How can you leverage this brain-based principle to increase long-term retention? Provide a summary of critical concepts learned during a virtual session and have participants review it right before bedtime. Give them an infographic or ezine, and ask them to spend ten minutes recalling what they learned earlier in the day. Ask them to restate what they recall in their own words. Text them a link to a short animated video that reviews key learning points and have them watch it before they take a nap or go to sleep.

NetSpeed Note

We created an animated video that reinforces the first three brain-based principles for engagement. You'll find it at our website: https://netspeedlearning.com/virtual-learning/brain-based-1-3/

Test and Retest

Most of us hate being tested. It's stressful and, if we fail, we feel bad about ourselves. We've grown up in educational systems that use testing to determine what grade we will get, what rewards we will receive, and, perhaps, even what college will accept us. Of course some graded testing should be required for people to demonstrate knowledge in a particular subject area. I'm glad my doctor went to medical school and passed her medical boards!

In corporate or organizational settings, testing might be required to be credentialed or to demonstrate compliance with laws and standards. We may believe that testing is a way to hold people accountable for their learning in the virtual classroom. The brain-based principle, **Test and Retest**, has nothing to do with proving that people were listening in class and therefore learned something. Recall that the memory formation process includes these four phases:

- Learn it
- Encode it
- Recall it
- Apply it

Testing and Retesting strengthens our ability to recall information so that we can successfully apply it on the job. The value of testing is that it gives learners the opportunity to remember what they learned, strengthening the neural connections that move information from short-term to long-term memory. I ask people who attend our Advanced Virtual Facilitator course how they can make testing fun (or at least make testing low-stress). Invariably, we talk about gamification. While I don't believe that gamification is the solution for all learning, I agree that gamification can be effective if it builds in rewards and positive reinforcement when people get correct answers.

There are a couple of online games I play regularly to relax. One is Candy Crush (I say that a little sheepishly). I find it interesting to ponder how Candy Crush works to keep me addictively playing. First, it does a masterful job of introducing new characters and challenges as I progress through levels. Once I learn how a particular aspect of the game functions, it repeats it in the next level, so I can build on what I have learned. When I start to get bored, it changes the rules, or adds a new element.

When I begin a new round or level in Candy Crush, I actually don't mind losing as I study the new rules and elements to see what I need to do to win. The game makers helpfully tell me how many times the average player takes to pass a round so they can move to the next level. I feel smugly superior at my obvious skill because I regularly beat that average.

There are rewards when I win (flashing lights, celebratory displays) and I can earn boosters that will help me win more quickly at the next level. In my happy brain, I am experiencing the release of the neurotransmitter dopamine[xxx], which is encoding what I need to do to repeat these rewards. What's happening to keep me coming back to Candy Crush is occurring in my brain's ventral system, and, most interesting to me, is directly tied to the learning process: learn it, encode it, recall it, and apply it.

What can we learn from a game like Candy Crush? For one thing, how are you rewarding the participation of people in your virtual classroom? It can be as simple as saying their name out loud when you comment on a point they've made in chat. You might set up competitions and games in real-time and reward the winners with a little hoopla. If your web conference platform has applause icons, you can use them to have people celebrate success. The brain-based principle **Test and Retest** encourages us to make testing a fun reinforcement experience. The emphasis should be on repeated recall to help transfer knowledge from short-term to long-term memory (moving it from the hippocampus to the neocortex). You're not trying to stump people. You're trying to reward them for successfully recalling what they have learned. When the time comes, they'll be ready to apply what they've learned, and ready to practice and master a new skill.

We design our virtual courses to include testing as a post-session reinforcement tool. Our leadership development and customer service training modules are delivered by a facilitator in the virtual classroom, and then followed by a testing experience in the NetSpeed Coach Learning Center. The questions we ask prompt the retrieval of key concepts as learners remember and apply their knowledge to simple case study questions. Correct answers are reinforced by a "coach" who appears in a video window. Incorrect answers are redirected. They can select an option to have their "coach" reteach a concept before they attempt to answer the question. As they move through the application, they are being scored. We encourage our clients to allow people to take the test as often as they need to, in order to achieve a passing score. After all, our hidden agenda here is reinforcement and reward.

Use It or Lose It

This final brain-based principle points to the importance of immediate use or application of new knowledge and skills. Scientists used to think that the brains of babies built millions of new neurons in infancy. Without use, those baby neurons would be pruned and die. By the time we hit adulthood, the collection of neurons we had left were what we got to work with from that point forward.

Neuroscientists now believe that our brains are constantly producing new neurons. However, according to the authors of "Use or Lose It: How Neurogenesis Keeps the Brain Fit for Learning,"[xxxi] there's good news and bad news about these new neurons:

> *"The good news is that new neurons are produced throughout adulthood. The bad news is that most of them do not survive. More than half of the new cells die within just a few weeks of being born."*

In other words, your role as a designer or facilitator in your virtual classroom is to stimulate new neurons and then help them survive! We get to be super heroes in this story.

What the research reveals is that your hippocampus produces new neurons as a result of being challenged in a learning experience. Those neurons have to become connected to a neural network with older neurons that have "been there, done that" if they are to survive beyond the learning experience. This is one reason why helping learners understand the context of new information through metaphor and storytelling is so important. We must strive to help their brains fire older neurons in synchrony with new neurons born during the learning process. And we've got about a week to strengthen those newbies through challenging application exercises before they start to die off.

What neuroscientists also know is that the more challenging the learning experience, the greater the neural connections those new neurons will make. Difficulty is essential for long-term memory. As our once new neurons are assimilated into our neural networks, it becomes easier to recall and apply what we've learned. If you are still presenting webinars as one-time events, with no follow-up application, you can be certain that new neurons are dying untimely deaths all around you.

The National Training Laboratory Institute for Applied Behavioral Science suggests that long-term retention is enhanced by the right teaching methods, as seen in this chart we use in our web workshops about brain-based learning:

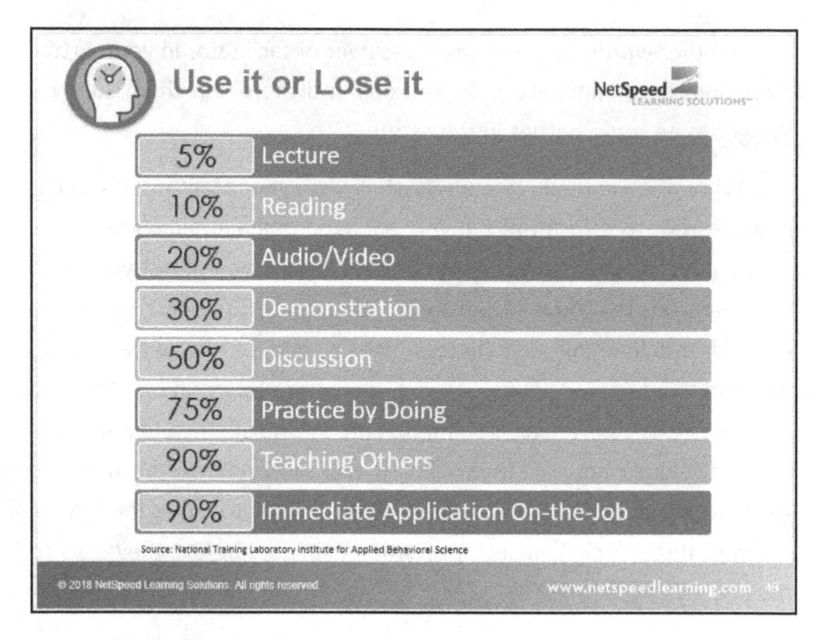

If you want minimal learning retention, lecture people (5%) or have them read articles (10%). You can improve it somewhat by adding demonstration (30%) and getting people involved in discussing the content (50%). But if you want learners to experience the survival of new neurons and the development of dynamic neural networks, they must practice by doing (75%), teaching others (90%), and immediately applying what they learned on the job (90%). Notice that these last three modalities are significantly more challenging than simply listening, watching, or discussing content.

Focus on helping participants in your virtual classroom integrate new neurons into the dynamic neural networks in their brains. Challenge them to practice, teach others, and apply what they've learned on the job. Build in course completion requirements to apply learning after a web workshop and then report back to their colleagues on their successes or failures. Reward on-the-job application of new skills, and ask people to write up their success cases to share with their colleagues and senior management.

The six brain-based principles for retention, combined with the six brain-based principles for engagement, ensure that the investments organizations make in developing people through virtual training are not wasted. I'm excited to continue exploring the cognitive principles that support building better online learning experiences. I invite you to join me in assuming responsibility for designing and delivering blended learning experiences that leverage the neuroscience of learning.

[xxiii] Hallowell, Edward. Shine: *Using Brain Science to Get the Best from your People*, Harvard Business Review Press, 2011.

[xxiv] Korb, Alex. *The Upward Spiral: Using neuroscience to reverse the course of depression, one small change at a time*, New Harbinger Publications, 2015.

[xxv] Medina, John. *Brain Rules*, Pear Press, 2014.

[xxvi] Bergland, Christopher, "Do Mirror Neurons Help Create Social Understanding?" Psychology Today, February 24, 2014.

[xxvii] Miller, G. A., "The Magical Number Seven, Plus or Minor Two," Psychological Review, 1956.

[xxviii] Sabourin, Randy, "Can Your Sales Training Hit a Curve Ball?" Training Magazine, February 2, 2017.

[xxix] Walker, Matthew, "The Role of Slow Wave Sleep in Memory Processing," Journal of Clinical Sleep

[xxx] Neuroscientifically Challenged Blog, "Know Your Brain: Reward System," January 16, 2015

[xxxi] Shors, T. J., Anderson, L.M., Curlik, D.M., and Nokia, S.M., "Use it or lose it: How neurogenesis keeps the brain fit for learning," NCBI, April 22, 2011.

Glossary of terms

Application Sharing Application sharing allows a presenter to share an application that is loaded on their computer. Other presenters (as long as they are given permission) can work on the same application at the same time.

Asynchronous learning Any learning event where interaction does not occur in real-time.

Chat Chat is an interaction tool available on most web conferencing platforms. Chat allows the facilitator and participants to communicate by text in real-time.

Collaboration Collaboration refers to the act of working together to achieve a goal, and often has the participants contributing independently from the facilitator.

eLearning eLearning (or electronic learning) is any computer-based training, especially via a network (like your corporate network or the Internet). This would include synchronous or asynchronous computer-based training or instructor-led training.

Interaction tool An interaction tool is a feature available on most web conferencing platforms that allows real-time interaction among the host, presenter, and participants. Examples of interaction tools are Chat, Poll, Annotation, and Whiteboard.

Interaction Interaction refers to verbal and written synchronous (two-way) communication, usually between the facilitator and the participants.

Polling The poll is an interaction tool available on most web conferencing platforms. A poll contains a question and one or more answers that can be selected by participants. The presenter is able to view the results as the poll question is being answered. They can then display the results, if desired.

Status Icons Status icons are a menu of tools available on most web conferencing platforms. They allow participants to signal their status or to make a request of the host or presenter. Examples of statuses include Raise Hand, Agree, Disagree, Step Away, Laughter, and Applause. Examples of requests include Speak Louder, Speak Softer, Speed Up, and Slow Down.

Streaming Video Streaming video is a one-way video transmission over a data network. Streaming video is displayed after a small amount of data has been transferred, unlike movie files, which are played only after the entire file has been downloaded.

Synchronous learning Any scheduled learning event where interaction occurs in real-time.

VoIP (Voice over Internet Protocol) VoIP provides streaming audio over the Internet, so that audio is heard over the learners' computer speakers. It's a cost-effective solution for large audiences. It only allows one person to be speaking at a time.

Whiteboard Whiteboard is an interaction tool available on most web conferencing platforms. The tool is used with annotation tool bars, similar to a Paint program, in that it allows a person to type text or draw graphics. Many web platform tools allow multiple people to draw, stamp, or type on the whiteboard simultaneously.

Appendix A: Virtual Facilitator Self-Assessment

This self-assessment is used for the homework from Chapter 1. It is a starting point to sound out your perspectives on the opportunities for successful learning transfer in the virtual classroom.

Perspective	Agree	Disagree
1. I believe that learners learn well in a virtual classroom.		
2. I find it difficult to facilitate virtually because I can't see the learners' body language or facial expressions.		
3. I want learners to engage with each other, not just with me.		
4. I think web conferencing is a one-way communication tool.		
5. I believe that learning in 1 to 2-hour chunks is preferable to long days in the classroom.		
6. Learners, in a virtual setting, need to be able to use a handout just like learners in a traditional classroom setting.		
7. Learners should receive a copy of the slide deck prior to the session so they can follow along.		
8. Most learners prefer to attend virtual classes because they can multitask while the facilitator lectures.		
9. Virtual training is a more efficient use of time because you can give learners the information quickly and they can get back to work.		
10. I strive to engage learners with interactive activities every 3 – 5 minutes in the virtual classroom.		

Appendix B: Interaction tools

The virtual facilitator should practice and master the interaction tools built into the web conferencing platform. Each tool helps to engage the learner in the active learning process.

Tool	Uses	Benefits
Chat	• Solicit learner input • Encourage collaboration	• Actively engages learners in discussion • Creates peer exchanges
Polling	• Check knowledge or experience • Stimulate interest • Set up lecture or discussion	• Provides instant feedback (and satisfaction) • Learners can compare their responses • Helps facilitator lead discussion and tailor lecture
Status Icons	• Quickly get input • Identify volunteers for exercises • See agreement or disagreement	• Participants can "vote" or respond when they may be uncomfortable using Chat • Opens the door for facilitator to call on learners to give examples
Streaming Video	• Streams video of Facilitator • Adds animation and interest	• Helps to establish rapport • Creates a sense of connection

Tool	Uses	Benefits
Whiteboard / Annotation	• Brainstorm and capture ideas • Encourage collaboration	• Allows facilitator to guide and record discussion visually • Encourages peers to share ideas
Application Sharing	• Share websites, your desktop, or documents • Can turn over control to specific participants	• Allows you to demonstrate steps or actions online • Gives a participant the opportunity to practice steps or actions while others observe
Breakout Rooms	• Have participants work in small groups	• Supports practice and feedback • Encourages quieter participants to participate verbally

Appendix C: Repurposing Samples

The following table describes three traditional classroom exercises, and how they can be repurposed for the virtual classroom. These exercises are used for the homework at the end of Chapter 6.

#	Traditional Classroom Exercise	Virtual Classroom Exercise
1	Participants break into three groups to discuss the benefits of delegating for the Organization, the Manager, or the Employee. Allow 5 minutes for discussion. Each group records ideas on a flipchart page and chooses a spokesperson to present the results to the class.	One Option: Assign people to groups 1, 2, or 3 based on last name (e.g. last name A-F is group 1). Open three chat pods: one for group 1 to chat about benefits for the Organization; one for group 2 to chat about benefits for the Manager; one for group 3 to chat about benefits for the Employee. Allow 3 minutes for chat discussion; then assign a spokesperson from each group to present the themes that emerged during the chat. (Unmute the spokesperson's phone line.)
2	Learners work in groups of four to review ten performance appraisal comments. They decide whether each comment belongs in the Hall of Shame (vague, subjective, and judgmental) or the Hall of Fame (objective, specific, and motivating). Each group prepares a flipchart with two columns: Hall of Shame and Hall of Fame and places each of the ten statements in one of the columns. Facilitator debriefs the exercise by identifying similarities	One Option: Present a set of polling questions using 5 – 7 performance appraisal comments. Have participants select either Hall of Shame or Hall of Fame for each poll. Display results poll by poll. When learners disagree, ask those who chose Hall of Fame or Shame to explain their reasons (Two options: Use the chat feature; or if using teleconferencing for audio, have

	and differences between the groups and asking for explanation or clarification.	learners unmute their phone lines to discuss their responses.) Invite participants to rewrite the Hall of Shame comments using Chat or Whiteboard.
3	Facilitator presents a six-step model to address performance issues. Facilitator may choose to model a positive example or show a video example. Learners complete a worksheet to prepare to role play or practice the model in an on-the-job situation. Learners break into groups of three to practice playing one of three roles: Manager, Employee, or Observer. In each round, participants rotate roles. Observers take notes and give feedback. Facilitator debriefs the exercise by soliciting examples of effective techniques or phrases.	One Option: Facilitator presents an overview of a six-step model to address performance issues. Facilitator and Host present a positive scenario applying the model. Facilitator guides learners step-by-step through a worksheet (sent in advance or downloaded from the platform), as learners prepare to practice the model in their own on-the-job situation. Facilitator asks for a volunteer to practice or role-play using the model. Learner plays the role of Manager; Host plays the role of Employee; the other participants observe and give feedback. The Facilitator coaches the "Manager" as he/she applies the model. Facilitator asks for feedback from observers in two chat pods: What Worked and What Suggestions Do You Have?

Appendix D: Using interaction tools for repurposing

Here are some tips showing how web platform tools can be used for repurposing. These tips are used for the homework at the end of Chapter 6.

Traditional Classroom	Virtual Classroom
Participant Introductions/ Mingles	1) Chat. Present an opening question. 2) Polling: Ask simple Yes/No question. 3) Icons: Ask for a show of hands.
Ice Breakers	Photo matrix: Whiteboard "one thing no one else knows about me."
Peer-to-Peer Discussions	1) Chat: Assign one question to each small group (4 – 5 peers). 2) Whiteboard: Share whiteboard; use group grid or matrix.
Self-Assessments	1) Handout: Complete questions in handout. 2) Then Polling: Share results with group.
Charting ideas on a flipchart	Whiteboard: Unmute phone lines and chart discussion.
Role-playing or practicing	Roleplay: Ask for two volunteers; unmute their phone lines. Have one play the role of manager; have the other play the employee, for example.
Modeling behavior	1) Host and Presenter: Present an effective interaction using a model or applying concepts. 2) Then Chat: Ask learners to give feedback.

Traditional Classroom	Virtual Classroom
Coaching participants	1) Polling: Present sample statement (or show video); ask whether it is effective or ineffective (Hall of Fame/Hall of Shame). 2) Then Chat: Ask how to improve the example.
Asking for questions from learners	Question & Answer Chat: Use separate chat or specific Q & A pod for questions from learners.
Asking open-ended questions	1) Chat: Pose interesting question and ask for input 2) Polling: Offer multiple choice responses; allow learners to see results. 3) Then Chat: Ask for additional clarification.
Asking closed questions	1) Icons: Use raised hands or green check to indicate agreement; red X to show disagreement. 2) Polling: Present closed question with Yes/No response; show results. 3) Then Call Out: Call on participants by name to explain their response.
Checking the pulse of the class	Icons: Ask for green check or red X regarding pace or content.
Presenting a video	Video: Send 1 – 2 minute video link prior to web session; have participants download it; or show it in web platform.
Checking for understanding	1) Icons: Use raised hands or green check. 2) Then Chat: Ask for application examples. 3) Then Audio: Call on volunteer with raised hand and unmute their phone line to share an example or say more.

Appendix E: NetSpeed Fast Tracks

People who participate in NetSpeed Learning's virtual courses utilize our NetSpeed Fast Tracks social learning website as a home base for communication, content sharing, microlearning, session introduction, polling, quizzes, class preparation, homework and post-class reinforcement.

The NetSpeed Fast Tracks environment, like other social media tools of its type, offers tools for self-paced content. Most of these tools contain a comments capability, which allows the class participants to offer feedback on each other's assignments. The following table displays a list of tools, with description and usage.

PodSpot
Description: A PodSpot is a short podcast, usually lasting between one and three minutes and presented in both an audio and visual format.
Usage: The instructor may explain a session or assignment using an audio PodSpot. Participants can link to PodSpots that contain course content.
Video
Description: Short video recording.
Usage: The instructor may explain a session or assignment using a video. Participants can link to a video that contain course content. The instructor may create a short welcome video from their desktop to introduce an online course.
Online Article
Description: Online articles are short, concise segments of information on relevant course topics. Within the article, you may find links to watch a video, download worksheets or other documents, or even read another online article.
Usage: The course facilitator may develop knols to contain course content. Participants may post course knols to respond to discussion questions or assignments.

Instructor and Class Blog
Description: Class blogs are a place to capture a thought, observation, or opinion. They are sometimes relevant for just a limited period of time. That's why newer blog entries (called blog posts) are displayed first, with older posts under them in reverse chronological order.
Usage: The class blog allows the course facilitators a forum to post informal thoughts, observations, opinions, and discussion questions. The blog is also where participants will share their insights and respond to specific discussion questions as part of their asynchronous assignments

NetSpeed Fast Tracks incorporates these tools into an integrated social learning platform that can even be brought into your organization. In this way, internal classes that might contain proprietary information can utilize a platform that is branded and secure for you.

Your company can maintain a level of control that doesn't exist if your content were to be distributed on YouTube or an Internet blog. There are built-in authoring levels so your administrators have control over what's posted.

Appendix F: Assessing a Virtual Training Program

The questions in the table below can help assess the effectiveness of a virtual training program. These were used for the homework at the end of Chapter 7.

Question	Yes	No
1. Do the learners have the opportunity to apply what they learned soon after training?		
2. Do they attend the online session with realistic expectations about the content and process?		
3. Have they identified at least one application prior to participating in the session?		
4. Does the manager prepare and support the learner before, during, and after the online session?		
5. Does the learner receive incentives, rewards, or encouragement to attend the session and apply what they learned?		
6. Do they attend the web classroom session close to a pressing need?		
7. Are they given tools and resources to apply the learning on the job?		

Appendix G: Delivering webinars to large numbers of participants

The ideal class size of the virtual classroom matches that of the traditional classroom: 15 to 20 people. The reason for this is that web workshops require engaged participants, and the more people, the less interaction is practical.

Some situations, like marketing webinars, require information transfer to a large number of people. The intent is less a transfer of learning than a transfer of enough facts that someone can make a decision.

With large audiences, expect lurkers (people who observe the presentation but do not participate by commenting in chat, answering polls, etc.). Here's a rough estimate derived from our experience at NetSpeed Learning.

Audience Size	Lurkers
15 – 20	Ideal size. No lurkers expected.
65	No lurkers if interaction tools are used effectively.
65 – 100	25% lurkers.
> 100	50% lurkers.

You should also be aware of bandwidth issues. The more attendees, the greater need to ensure that you have enough bandwidth for streaming video, for 100 people chatting, etc. Teleconferencing is an expensive option. Chances are good you'll be using VoIP. It's a cost-effective audio solution for large audiences. Though listeners technically may be able to speak through their computer microphone (if you've enabled two-way VoIP), you'll probably keep them muted.

How do you know in advance how your platform will react to bandwidth constraints of large audiences? You can only know through experience with your own platform and technology environment. Here are some things that we've learned about bandwidth and VoIP:

- The big bandwidth hog is streaming video. It may be difficult to use if there are more than 500 people on the call using VoIP.
- If you do try to stream video and voice, the host and presenter should close every other application that might be interfering with your bandwidth. If your learners complain of a bumpy ride, then replace the video with a picture of the facilitator.
- If only a few learners are experiencing issues, have them close other applications on their desktops that may be interfering with performance.
- Don't forget that if a participant doesn't have speakers or headphones connected to their computer, they won't hear the "voice" in VoIP.
- With VoIP, usually only the person speaking has control; participants can't talk over each other.
- You probably won't want to use breakout rooms with these larger groups. Most web conference platforms have limits to the number of people that can be placed in breakout rooms. Limit your presentation and interactivity to PowerPoint slides, polling, chatting, and Q&A pods.

Give some thought on how to use interaction tools intelligently. For example, in a typical web workshop you will often pose a question using a Poll, then follow up the results with Chat. In a marketing webinar, offering a poll is still a good idea – it keeps people focused and (at least theoretically) takes as little time for 100 people as it does for 10. Following the poll, you can request more feedback to their poll response via Chat, but 100 responses fly by too quickly to read.

This scenario might work in some situations, but not in others. For example, if you asked your audience what cities they were attending from, it might be fun to scroll down the long list of responses. But if you asked them for a substantive reply and could only comment on a small percentage of them, you will be in danger of losing your audience. The next time you ask them a question, they'll think "What's the use of answering?"

Another strategy might be to direct a chat to a smaller segment of your audience. You might divide them into three or four groups by last name, for example.

The takeaway is to be aware of the limitations of interaction tools for large audiences, and be creative in their use.

Appendix H: Creating a Risk Management Plan

When working with technology, it often seems that if something can go wrong, it will go wrong. To give yourself peace of mind, it's helpful to have a risk management plan, as was discussed in Chapter 8.

Standard risk management teaches us to anticipate the things that can go wrong (the "threat"), do what we can to prevent them from going wrong ("avoid"), and mitigate their negative effect ("mitigate") if they do go wrong.

On the next page is a sample Risk Management plan. Not only is this helpful to think about beforehand, but you or your host can also keep a print-out by your desks during the webinar.

The first thing to do is to brainstorm the threats. The first column of the table lists some mishaps that were mentioned in Chapter 1. Next, come up with ways that you can avoid the threat (in the second column), as well as ways to decrease the consequence if the threat does occur (in the third column).

If you want to get more sophisticated and prioritize the risks (so you know which ones to spend the most time on) use the Risk Value Analysis (on the page following the Risk Management Plan). The values in the second and third columns are percentages of the likelihood of the threat occurring and the impact of the threat's repercussions, respectively. Risk Value is simply the likelihood multiplied by the impact.

Risk Management Plan

Threat	To avoid	To mitigate
Presenter can't log in (due to # of seats being maxed out)	Don't open the webinar to participants until the presenter is logged in	Take over the host's computer
Host exits session prematurely (due to either user or technical error)	Always have both the host and presenter log in as Hosts	Quickly create a new session, email it to participants, and ask them to log in to the new session
Presenter's audio transmission is jerky	Give instructions during housekeeping to close competing programs	Use Q&A pod, have the host advise
Participant joins breakout room but has no audio	Have participant log in correctly (instead of calling in to the teleconference bridge, log into the web conference and have the system call back)	Have participant hang up and have the system call back
Participant unable to return from a breakout session	Train users before the webinar to log in to the web conference event first, and then have the system call them.	Before sending participants to break out rooms, check the Attendee list to be sure no one has dialed in directly.

Risk Value Analysis

Threat	Likelihood	Repercussion	Risk Value (*l* X *r*)
Presenter can't log in (due to # of seats being maxed out)	.10	1.00	.10
Host exits session prematurely (due to either user or technical error)	.10	1.00	.10
Presenter's audio transmission is jerky	.90	.20	.18
Participant joins breakout room but has no audio	.25	.90	.225
Participant unable to return from a breakout session	.50	1.00	.50

In the above table, the threats are listed in increasing order of risk value. The repercussions of a presenter not being able to log in are high, but because the likelihood is so low, the risk value is also low. The values for the presenter's audio being jerky are reversed: the likelihood is high but the impact is low. However, the repercussions of a participant not being able to return from a breakout session are high, and the likelihood is fairly high. The combination produces the highest risk value in this chart. I'd recommend practicing your mitigation strategy for this one.

Appendix I: Logistics checklist

Here's a checklist you can use to accompany what you learned in Chapter 9. It includes some final steps to walk through before performing your first virtual facilitation class.

Preparation

1. Set up meeting/event in web platform
2. Send invitations
3. Schedule Dry Run
4. Prepare Power Point for web platform
    a. Upload or convert to acceptable format, if appropriate
    b. Check converted files, slide by slide
5. Prepare polls, chats, whiteboards, and video for web platform
6. Print thumbnail copies of slides for Host and Presenter, if needed

Dry Run

1. Ensure assets are loaded and ready to go
2. Walk through web workshop with Host
3. Determine roles/responsibilities for Host and Presenter
    a. Determine who opens and closes the meeting event
    b. Decide who advances slides, opens polls, etc.
4. Practice passing the baton from section to section
5. Check video/audio settings
    a. If using VoIP, test headset
    b. If using teleconference, test headset and test bridge number
6. Test recording feature, if needed

Revisions

1. Make corrections to slides, if needed
2. Upload or convert slides to acceptable format
3. Adjust interactive exercises, as needed

Dress Rehearsal (with pilot participants)

1. Invite test group of 5 – 10 participants
2. Present web workshop to test group, using all interaction tools

3. Test recording features

Delivery

1. Deliver final web workshop to actual participants

Debrief

1. Debrief session for lessons learned
2. Document best practices

Appendix J: Design an interactive, collaborative learning experience

These questions and steps pull together all the information you've learned in this book to design and deliver an interactive, collaborative learning experience.

Preparation questions

- What are the learning objectives for this session?
- How can I engage the learners throughout the session?
- How can I make lectures short (3 – 5 minutes) and interesting?
- For every 60-minute session, where will I include 10 – 20 interactions using chat, polling, whiteboards, annotation tools, and status icons?
- At what points will I encourage peer-to-peer collaboration?
- At what times will I call on learners to give specific examples or responses?
- How will I ensure that every person in the session participates?
- What pre-class assignments will engage their interest and commitment to learning about this topic? What relevant examples or challenges will I request in advance?
- How can I interact with a host or co-facilitator to stimulate interest and capture attention?
- How will I encourage learning transfer?

Prior to the web session

- Send an introductory email telling participants what to expect during the session
- Solicit relevant examples and problems

- Send handout pages; give an easy assignment to be completed before class
- Do not send out slide deck (even though learners may request it)

At session opening

- Review learning objectives
- Go over housekeeping issues (e.g., mute your phone line; close other programs)
- Ask for their expectations
- Practice interaction and collaboration quickly
- Chat (say "hello" to everyone)
- Poll ("How much experience with this topic?")
- Raise Hand or Thumbs Up (pose a simple question)
- Whiteboard
- Describe level of participation expected
- Tell them you will be calling on people by name
- Ask them to suspend multitasking

Following the session

- Send slide deck for reference
- Send post-session survey requesting examples of on-the-job application
- Tie certificates of course completion to active participation during session

Appendix K: Resources for the virtual trainer

Adobe

www.adobe.com

Take a look at Adobe Acrobat Connect if you want to use one of the most flexible web conference platforms on the market today. And while you're there, explore Adobe Presenter and Adobe Captivate, products that support more sophisticated PowerPoint slide presentations within Connect.

Logitech

www.logitech.com

Logitech is one of the major suppliers of webcams. For high quality video and audio, at a reasonable price, add a Logitech webcam. You'll be able to add streaming video to your web event. They also make Internet headsets which you'll need for VoIP.

OttLite

http://www.ottlite.com/

A great source of full spectrum lights for use in lighting your face when you are on camera. Buy a desktop lamp and set it next to your laptop or monitor.

Plantronics

www.plantronics.com

Don't skimp on your telephone headset. We recommend a wireless headset from Plantronics for good sound quality and hands-free web conference audio.

Zoom

www.zoom.us

Check out Zoom for easy-to-use web and video conferencing. It's user-friendly and may be the perfect first step for the novice virtual trainer or facilitator.

Deposit Photos

www.depositphotos.com

Once you've spent some time at this website, you'll never settle for boring slides again. Royalty-free art, photos, illustrations and more are available to very reasonable prices. A great resource if you're ready to move beyond your bullet point slides.

Random.org

www.random.org/dice

Sometimes you just want to randomly select questions, problems, challenges, or groups virtually. So show them the random dice roller at this website during the web event. It's silly but attention-getting.

Resources available at NetSpeed Learning Solutions

Virtual Facilitator Trainer Certification

http://www.netspeedlearning.com/interactive/vftc/

Designed for experienced classroom trainers who want to become experts at designing and delivering engaging, highly interactive web conference training, this course is delivered through a blend of facilitated webinars and self-paced online content. Assignments, discussions, and exercises are completed each week on NetSpeed Fast Tracks™. At the end of the course, trainers will prove their virtual facilitation skills by designing and delivering a 20-minute webinar to their course colleagues.

Web Conference Essentials™

http://www.netspeedlearning.com/interactive/essentials/

Designed to give participants an experiential overview of the best practices used in engaging, interactive virtual learning, this program includes a series of three interactive 60-90 minute demonstration webinars. Participants will experience a variety of facilitation techniques and approaches, complete take away exercise samples, and have the chance to apply their new concepts to an on-the-job project. A Master Trainer will observe their next webinar delivery and provide written feedback on its effectiveness, if requested.

NetSpeed Fast Tracks

http://www.netspeedlearning.com/fasttracks/

Leverage emerging technologies using powerful tools such as podcasts, videos, blogs, and microlearning, with NetSpeed Fast Tracks™, our customizable, online learning platform that allows participants to engage in self-paced, learning activities in a virtual, collaborative learning environment.

NetSpeed Leadership

http://www.netspeedlearning.com/leadership/

NetSpeed Leadership is designed to meet the learning needs of managers, supervisors and individuals in fast-paced organizations. Using interactive instruction (for both the face-to-face and virtual classrooms) coupled with powerful, easy-to-use microlearning tools for online reinforcement and for measuring impact, the NetSpeed Leadership system successfully ensures learning transfer, holds participants accountable and empowers them to apply new skills on the job.

Managing Workplace Conflict

http://www.netspeedlearning.com/managing-conflict/

The Managing Workplace Conflict training program develops the skills, behaviors and practices that allow employees to communicate, collaborate, and resolve conflict.

Based on the book *Managing Workplace Conflict*, the program brings the principles and strategies from the book to life using a combination of discussions, role plays and case studies. Participants interact with the facilitator and with each other to connect the course material with their personal experiences. Skill-building scenarios, videos depicting the different behavior types, and audio clips contribute to an immersive experience that models successful engagement to resolve difficult scenarios.

Blazing Service

http://www.netspeedlearning.com/blazingservice/

Blazing Service™ develops customer service skills, helps increase customer satisfaction, and improves customer retention by combining the best of classroom instruction (face-to-face or virtual) with easy-to-use, web-based reinforcement tools. Designed to meet the learning needs of customer service providers in high customer-contact organizations, Blazing Service helps employees quickly grasp and apply proven interpersonal and problem-solving techniques to ignite great customer service in your organization.

NetSpeed Virtual Learning

http://www.netspeedlearning.com/virtual-learning

We'll work with your instructional designers to repurpose your existing classroom training content for high-impact virtual delivery, and blend synchronous, facilitated web delivery with asynchronous pre- and post-webinar microlearning tools to produce programs that get results. Using your preferred web conference platform, we'll help you develop and launch high-quality, virtual learning. If you choose, we can blend your synchronous programs with NetSpeed Fast Tracks™ to add the power social, collaborative learning to your programs. We're experts at blending synchronous classroom and asynchronous microlearning elements to ensure powerful collaborative learning that sticks.

About the Author

Back in 2000, Cynthia Clay founded NetSpeed Learning Solutions with the belief that technology would not just revolutionize the way training was delivered, but that it also held the potential to dramatically enhance and extend the learning experience itself. With 23 years in training and management development and a past career that includes training, coaching and mentoring thousands of employees and trainers at hundreds of companies, Ms. Clay knew that technology's advantage of rapid delivery and virtual access, while powerful, was just the beginning.

With a passion for using technology in the service of learning, Ms. Clay has channeled her vision into creating a company that specializes in combining brain-based learning principles, meaningful content, and engaging delivery with the dynamic, media-rich features of emerging new technologies. Under her leadership, NetSpeed Learning Solutions is an industry leader in the development of innovative, successful blended and virtual learning programs, incorporating social learning and microlearning

Ms. Clay is a recognized thought leader and expert in the field of blended and virtual learning strategies. She is also a sought-after speaker at industry conferences.

Ms. Clay is an active member of the local and national community. Her work has included participation in the following organizations:

- Member, Women Presidents Organization
- Member, Chairman of the Board of Directors, & Sponsor of the Women's Business Exchange

- Board of Trustees, Center for Spiritual Living (Seattle)
- 20-Year Member, ATD (Association for Talent Development)
- 10-Year Member, ISA (The Association of Learning Providers)

Index

Bring *Great Webinars* to your Organization!

Order via the Web

Purchase individual copies for $34.95 each plus tax and shipping: https://netspeedlearning.com/greatwebinars/

Order Copies for Your Organization

For larger orders, call us at 206-517-5271 (Toll-Free in US / Canada): 877-517-5271

2-10 copies	$32.00 each
11-100 copies	$25.00 each
Over 100	Please call us for pricing.

Customize This Book

We can tailor this book for your organization, including featuring your logo and/or branding on the cover and even customizing the content to reflect your learning environment. Please call us for more information.